Internet of Things

Technology, Communications and Computing

Series Editors

The series Internet of Things - Technologies, Communications and Computing publishes new developments and advances in the various areas of the different facets of the Internet of Things. The intent is to cover technology (smart devices, wireless sensors, systems), communications (networks and protocols) and computing (theory, middleware and applications) of the Internet of Things, as embedded in the fields of engineering, computer science, life sciences, as well as the methodologies behind them. The series contains monographs, lecture notes and edited volumes in the Internet of Things research and development area, spanning the areas of wireless sensor networks, autonomic networking, network protocol, agent-based computing, artificial intelligence, self organizing systems, multi-sensor data fusion, smart objects, and hybrid intelligent systems.

Indexing: *Internet of Things* is covered by Scopus and Ei-Compendex **

Kannimuthu Subramaniyam · Marcin Paprzycki
Editors

Smart Cities and the Internet of Things

The Evolution of Cities

Springer

Editors
Kannimuthu Subramaniyam
Karpagam College of Engineering
Coimbatore, Tamil Nadu, India

Marcin Paprzycki
Systems Research Institute Polish Academy of Sciences
Warsaw, Poland

ISSN 2199-1073 ISSN 2199-1081 (electronic)
Internet of Things
ISBN 978-3-032-08289-3 ISBN 978-3-032-08290-9 (eBook)
https://doi.org/10.1007/978-3-032-08290-9

This Springer imprint is published by the registered company Springer Nature Switzerland AG
The registered company address is: Gewerbestrasse 11, 6330 Cham, Switzerland

Committee Members

General Chairs

Prof. Yuexia Zhang, Beijing Information Science and Technology University, China
Prof. Gaoru Zhu, Beijing Jiaotong University, China

Technical Program Committee Chairs

Prof. Mouquan Shen, Nanjing Tech University, China
Prof. Hongke Xu, Chang'an University, China
Prof. Dr. Seyed Mohammadreza Ghadiri, SIMI (Swiss Information and Management Institute), Switzerland
Prof. Kannimuthu Subramanian, Karpagam College of Engineering, India

Organizing Committee Chairs

Prof. Xin Yang, Beijing Jiaotong University, China
Assoc. Prof. Gaohong She, Beijing Jiaotong University, China
Assoc. Prof. Han Wang, Nantong University, China

Publication Chairs

Prof. Gaoru Zhu, Beijing Jiaotong University, China
Assoc. Prof. Marcin Paprzycki, Polish Academy of Sciences, Poland

Committee Members

Prof. Chun Zhang, Beijing Jiaotong University, China
Professor-level Senior Engineer, Chunyi Yuan, Ministry of Transport Planning and Research Institute, China
Prof. Md. Sipon Miah, Islamic University, Bangladesh
Assoc. Prof. Thomas Lee, University of Illinois at Chicago, USA
Assoc. Prof. Nana Cui, Beijing Jiaotong University, China
Assoc. Prof. Lu Chen, Beijing Jiaotong University, China
Assoc. Prof. Nur Khairiel Anuar, Universiti Utara Malaysia, Malaysia
Assoc. Prof. Gajendra Sharma, Kathmandu University, Nepal
Assoc. Prof. Muath Obaidat, City University of New York, USA
Assoc. Prof. Goi Bok Min, Universiti Tunku Abdul Rahman, Malaysia
Senior Lecturer, Xiaochun Cheng, Swansea University, UK
Senior Lecturer, Farkhana binti Muchtar, Universiti Teknologi Malaysia, Malaysia

Preface

The 2025 5th International Conference on Smart City Engineering and Public Transportation (SCEPT 2025) was held successfully in Beijing, China, from March 28 to 30, 2025. As an established international forum, SCEPT continues to promote the advancement of research and practice at the intersection of smart city development and modern transportation systems. The 2025 edition attracted over 100 attendees, including researchers, engineers, and policy practitioners from academia, industry, and government sectors. With a total of 80 paper submissions and 23 accepted for inclusion in the proceedings, the conference maintained a rigorous peer-review process, fostering a high standard of academic contribution and scholarly exchange.

This year's conference featured four distinguished keynote speakers who provided profound insights into critical issues in smart city infrastructure and transportation resilience. Professor Mouquan Shen from Nanjing Tech University presented his latest research on next-generation communication mechanisms for intelligent transportation systems, highlighting innovations in vehicular network protocols and real-time control. Professor Seyed Mohammadreza Ghadiri from the Swiss Information and Management Institute offered a cross-national perspective on the digital transformation of smart cities in developing economies, addressing urbanization challenges and ICT-driven policy frameworks. Professor Xin Yang of Beijing Jiaotong University shared a systems-level analysis of the resilience of urban rail transit networks, drawing from recent developments in complexity modeling and multiscale simulation. Professor Gaoru Zhu, also from Beijing Jiaotong University, discussed spatial suitability evaluation and optimization strategies for transportation infrastructure within the framework of national territorial planning.

Beyond the keynote presentations, SCEPT 2025 included several parallel oral sessions that showcased a wide spectrum of innovative studies and engineering applications. The presentations reflected the conference's multidisciplinary scope, which spans topics such as intelligent vehicle control systems, traffic flow modeling, multimodal logistics networks, and port and inland waterway traffic coordination. In the realm of smart city technologies, contributions addressed data-driven urban management, city-scale simulation, signal transmission and noise reduction in urban communication networks, as well as the integration of cloud computing, Internet of Things (IoT), and green communication technologies for sustainable urban development.

The scholarly exchanges at SCEPT 2025 underscored the increasingly intertwined nature of urban infrastructure planning, intelligent transportation, and information systems engineering. By bringing together diverse perspectives from engineering, information science, and public policy, the conference served as a valuable platform for forging interdisciplinary collaborations and identifying strategic directions for future urban innovation.

We are deeply grateful to all contributors, including authors, keynote speakers, and reviewers, whose rigorous work and academic dedication have shaped the quality and richness of this volume. We also thank the organizing committee for their professional efforts in ensuring the smooth execution of the conference. We hope that the insights and results presented herein will inspire continued research in smart city engineering and intelligent transportation and provide meaningful references for scholars and practitioners working to build smarter, more resilient urban futures.

SCEPT 2025 Conference Committee

Contents

EMD-LSTM-Based Prediction of Passenger Flow in and Out of Subway Stations 1
Jiahui Yang, Zhe Ji, and Fengwu Wang

Research and Application of Mobile Phone Signaling Data in Urban Comprehensive Traffic Planning 13
Chunrong Xie, Tao Yang, Yanzhi Pang, and Xiali Cao

Prediction of Taxi-Out Time Based on LSTM-XGBoost 25
Guoxin Zhang, Guan Lian, Benxiao Lou, and Wenyong Li

Research on China Freight Volume Forecasting Based on LSTM 39
Quanli Mo, Liwei Li, Kugan Huang, Yanfang Pan, Shiyu Wang, Bote Liu, and Guobin Gu

Urban Flood Simulation and Resilience Assessment Based on Cellular Automata and System Performance 49
Yuelei Wu, Rupeng Zhao, Qingguo Li, and Yang Chen

Research on the Scale Calculation Model of Port Logistics Park 59
Shiwen Sun, Peng Cai, and Jinxiu Fu

Intelligent Connected Vehicles Actively Coordinate Collision-Avoidance Control Based on Lane-Change Trajectory Prediction of Sidecars 69
Dengcheng Liu, Yifan Hu, Zhipeng Sun, Zhouping Zhang, Ruiyang Xiong, and Yunfeng Chen

A CRITIC-Cloud Model Coupling Framework for Dynamic Evaluation and Optimization of Urban Rail Transit Emergency Plans 79
Chaochen Wang, Wenkai Xu, Jun He, Xinhuan Zhang, and Jinhong Wu

Highway Traffic Flow Congestion Warning Model Based on the Weighted Linear Extrapolation Method 93
Chun Bao, Guoyan Che, Jianqiu Chen, Yanzhi Pang, Benxiao Lou, Mingzhe Chen, Fanglan Tan, Weijie Chen, and Guobin Gu

Study on Highway Asphalt Pavement Performance Decay Model in Xinjiang Based on Holt's Exponential Smoothing Prediction Method 105
Chunmei Liu, Mingyuan Chang, Guobin Zhang, Yihua Wang, and Xiaomin Dai

Research on Electric Vehicle Scheduling Methods Considering the Uncertainty of Users' Scheduling Willingness 117
Ran Lyu, Mingxing Guo, Li Lan, Yong Zhang, Su Wang, and Zikang Ding

Multisource Information Fusion-Based Bus Origin–Destination Prediction Model 129
Zeyu Wen, Mingquan Shi, Yi Li, and Wenyuan Wu

Application and Empowerment Path of Digital Twin Technology in Automobile Manufacturing Industry 141
Yongkang Liang, Kuo Song, Shun Wang, and Jingli Han

Design and Realization of LoRa-Based Low-Power Smart Ranch Environment Monitoring System 151
Jiajie Tao and Yongfei Lu

A New Smart Home System Based on IoT from 5G 167
Zhaoye Li, Jiali Quan, and Mengying Liu

Design and Implementation of Intelligent Kitchen Environment Monitoring System 177
Yuting Jiang

Design of Intelligent Operation and Maintenance System of Ticket Sale and Inspection Equipment of Urban Rail Transit Station Based on PHM Technology 187
Haitao Deng, Xiali Cao, Hao Li, and Yanlan Meng

The Design of a Virtual–Real Hybrid Simulation System for Airborne Weather Radar 199
Shiwei Zhao, Menglong Shao, Feilong Huang, and Xibo Cao

Research and Development of a BIM+IoT--Based Digital Command and Decision Platform for Construction Engineering 211
Junyi Li, Xiaodan Ma, Hang Yu, Yansong Wang, Xuejuan Dou, Pengzhou Wang, and Yang Wang

Design and Implementation of the Low-Altitude Intelligent Traffic Management System for Urban Environments 229
Xiaoxia Zhou, Tao Tang, Xiang Zou, Xi Fu, and Xinyi Yang

Design of Underground Pipe Gallery Monitoring and Early Warning System Based on Digital Twin Technology 243
Yingxia Li and Weiying Hao

eVTOL–Ground Vehicle Dynamic Coordination Optimization Algorithm Based on Model Predictive Control 253
Mutong Ji, Sen Liu, and Tianqi Yuan

Vehicle Environmental Audio Recognition Based on Convolutional Neural Network and Time–Frequency Analysis 267
Zeyu Li

EMD-LSTM-Based Prediction of Passenger Flow in and Out of Subway Stations

Jiahui Yang, Zhe Ji, and Fengwu Wang

Abstract In order to improve the accuracy of subway traffic prediction, an empirical mode decomposition (EMD) and long short-term memory (LSTM) model is proposed, and the model is verified by taking Xinjiekou Station of the Nanjing Subway as an example. The study shows that the EMD-LSTM model has the lowest root-mean-square error (RMSE) and mean absolute percentage error (MAPE) compared with the autoregressive integrated moving average (ARIMA) model and the LSTM model, which indicates that the model has high prediction accuracy and can be well applied to the prediction of subway passenger flow.

Keywords Passenger flow · Empirical mode decomposition (EMD) · Long short-term memory (LSTM)

1 Introduction

At present, the construction of public transportation in China has entered a period of rapid development, among which the subway occupies a large proportion in urban transportation with the advantages of high efficiency and punctuality, and gradually becomes the backbone of the urban transportation system [1]. According to the relevant data [2], as of January 1, 2019, a total of 35 cities in mainland China had opened and started the operation of metro, with the total mileage of metro operation reaching 5027.36 km, and the cumulative annual passenger volume of the country reached 23.71 billion passenger trips, an increase of 12.2% compared with 2018, among which the average daily passenger flow in large cities can reach 10 million. With the increase of the passenger flow in and out of subway stations (hereinafter referred to as "passenger flow"), some sections and stations of the subway lines show the characteristics of insufficient operational capacity, and the high density and loading rate of subway station platforms during peak hours have become the norm in large cities. The increase in passenger traffic seriously affects the quality of

J. Yang · Z. Ji · F. Wang (✉)
Navigation College, Dalian Maritime University, Dalian, China
e-mail: 498220772@qq.com

K. Subramaniyam and M. Paprzycki (eds.), *Smart Cities and the Internet of Things*, Internet of Things, https://doi.org/10.1007/978-3-032-08290-9_1

service for passengers on the one hand and exacerbates the unevenness and potential unsafe factors in the operation of subway lines on the other. For example, on July 5, 2011, a stampede occurred at the Zoo Station of Beijing Subway Line 4, resulting in the death of a 12-year-old boy, two serious injuries, and 28 minor injuries [3]. The reason for this is mainly due to the fact that the subway operates in a more closed environment underground and has a high concentration of passengers, and these unfavorable conditions can limit the evacuation of people in the event of an unsafe accident [4]. Therefore, with a scientific and reliable model to predict the passenger flow, it is possible to know its changes in advance, based on which the prediction results can provide a basis for passengers to choose their travel time; it can also provide a reference for the metro-related departments to optimize train scheduling, reasonably allocate staff on duty, and deploy security measures to further ensure the safety of passengers' travel. In view of this, it is very necessary to forecast passenger flow.

In 1995, Helbing et al. [5] proposed the "social force" model for the first time based on the analysis of passenger flow characteristics of subway passengers, which has been unanimously accepted in the study of passenger flow organization planning and has been widely applied in the study of passenger flow. In 2004, Daamen [6] established a passenger travel route selection model based on the consideration of infrastructure services in metro stations and metro information services and also analyzed and summarized the passenger travel law, which laid the foundation for the subsequent study of passenger flow law. In 2006, Zhang [7] conducted a case study on urban rail passenger flow in Xi'an from the aspects of passenger flow change law and passenger flow planning. In 2006, Zhang [7] conducted an example verification of the passenger flow in Xi'an City, established a corresponding prediction model, and analyzed the main reasons for the large error of urban rail passenger flow prediction results in combination with the factors affecting passenger flow; the conclusion provided a reference for the subsequent research on passenger flow prediction model. In 2011, Zhang and Chen [8] analyzed and researched the passenger flow change law in metro stations and proposed optimization measures for passenger flow organization from the aspects of passenger flow prediction and passenger flow organization.

It can be seen from the study of related literature that there are three main characteristics of passenger flow changes as follows:

(1) Passenger flow is a sequence of values that changes with time, and the corresponding passenger flow on each time period is closely related to the previous time periods.
(2) Passenger flow generally has a cyclical pattern but also has nonlinear and unstable characteristics.
(3) According to the operating date of the subway, passenger flow can be divided into normal passenger flow and non-normal passenger flow. The normal passenger flow refers to the passenger flow under normal conditions on normal days (weekdays); the non-normal passenger flow refers to the passenger flow under

special circumstances such as weekends and large events, and there are obvious differences between the two kinds of passenger flow variations.

Based on the characteristics of passenger flow changes, many corresponding forecasting methods have been developed. The current forecasting methods are mainly divided into parametric and nonparametric methods [9]. Parametric methods mainly include time series methods, which are based on statistical principles and seek the inherent patterns of historical data in order to estimate the future passenger flow data at a given observation point on a specific time period—for example, Markov model, Kalman filter, autoregressive integrated moving average (ARIMA) model, etc. Although these methods have the characteristics of simple model and high operability, the simple mathematical statistics cannot reflect the nonlinear relationship existing in the passenger flow data. Therefore, the fitting effect for the nonlinear data of passenger flow is not ideal. Nonparametric methods mainly include neural network model. Although the traditional nonparametric methods can make up for the shortcomings of the parametric methods in fitting the nonlinear data of passenger traffic, they also have the defects of long training time, low timeliness, and insufficient model depth.

In recent years, with the rise of deep learning, long short-term memory (LSTM) models have been developed, which are among the nonparametric methods, frequently used in traffic flow and passenger flow prediction because they are more suitable for processing time-series data, with the advantages of short training time, strong nonlinear fit, and high prediction accuracy. However, the LSTM model is more demanding for the input data, and the unstable nature of the original sequence will have a greater impact on the prediction accuracy of the model.

In the traditional passenger flow prediction methods, most scholars ignore the characteristic of the original sequence itself being unstable. This characteristic is mainly caused by the noise from the sample, and it will have a large impact on the model accuracy, so it is especially important to deal with the sample noise.

Empirical mode decomposition (EMD) is a signal analysis method that can handle nonlinear and nonstationary data and is widely used in time series for data preprocessing. It is able to decompose the sample into a finite number of intrinsic mode function (IMF) components and one residual and then achieve sample noise reduction by processing each component separately.

In this paper, we combine EMD and LSTM models to propose an EMD-LSTM passenger flow prediction model and use Nanjing Metro Xinjiekou Station as an example for prediction. It is expected that the prediction results can provide some theoretical support for passenger travel and the optimization of passenger flow organization by metro-related departments.

2 Algorithm Introduction and Model Construction

2.1 Empirical Mode Decomposition

EMD is a signal analysis method for nonlinear, unsteady data, which is fundamentally different from a priori harmonic and wavelet basis functions. EMD does not require any predetermined basis functions and identifies all vibration modes contained in the signal based on the time-scale characteristics of the data itself, decomposing them into a finite number of IMF components and one residual.

The IMF components generated by EMD should satisfy the following two requirements:

(1) The number of extreme points of the IMF must be equal to the number of zero points or differ by at most one.
(2) The mean value of the envelope defined by the local maximum and the mean value of the envelope defined by the local minimum should be zero at any moment point of the IMF.

Let there be m sets of original passenger flow data and z_t ($t = 1, 2, \cdots$, m) be the passenger flow time series, whose EMD decomposition process is as follows:

Step 1: Identify all maximum and minimum value points based on the original series.
Step 2: Apply the three-time spline curve to fit the envelopes of the upper and lower extreme points, denoted as $e_{\max}$ and $e_{\min}$, respectively, and find the average of the envelopes of the upper and lower extreme points u_t.
Step 3: Calculate the difference between z_t and u_t and integrate them into a new sequence v_t.
Step 4: Determine whether v_t satisfies the IMF condition. If the IMF condition is not satisfied, repeat steps 1–3 with v_t instead of $\mathrm{z_t}$ z_t until v_t satisfies the criterion; then V_t v_t is IMF, denoted as v_i.
Step 5: Repeat steps 1–4 k times until the residual becomes a monotonic function or less than the preset error; then the EMD decomposition is terminated. The residuals are noted as R_{es}.

The above steps can decompose the original signal into n IMF components and one residual.

$$z_t = \sum_{i=1}^{n} v_i + R_{\text{es}} \tag{1}$$

2.2 Long Short-Term Memory

The LSTM model is an improvement on the recurrent neural network (RNN) model and therefore belongs to a type of RNN network. It mainly circumvents the gradient explosion and gradient disappearance problems in the standard RNN network structure, so the training speed will also be faster. The network structure mainly consists of a cell state used to store long-term memory and input gates, forgetting gates, and output gates. The cell state is used to store the information of historical memory and present memory, and the three gates are used to control the operations of adding and deleting information.

The structure of the LSTM model is shown in Fig. 1, and the calculation steps for a single LSTM neuron cell (hereinafter referred to as "cell") are as follows.

Step 1: The LSTM model first determines how much information needs to be retained in the cell state C_(t–1) at the previous moment through the forgetting gate, i.e.,

$$f_t = \sigma(W_f \cdot [h_{t-1}, x_t] + b_f) \tag{2}$$

where f_t is the output value of the forgetting gate and σ is the activation function Sigmoid, i.e., the output result is mapped to the interval [0, 1]; W_f is the weight of the forgetting gate; h_(t–1) is the output value of the network at the previous moment; x_t is the output value of the network; and b_f is the bias of the forgetting gate.

Step 2: The input gate mainly decides how many components of the input x_t are retained in the cell state C_t. The current cell state (C_t) and the previous moment cell state C_(t–1) are combined into the new cell state C_t output, i.e.,

$$i_t = \sigma(W_i \cdot [h_{t-1}, x_t] + b_i) \tag{3}$$

Fig. 1 Structure of neurons in LSTM model

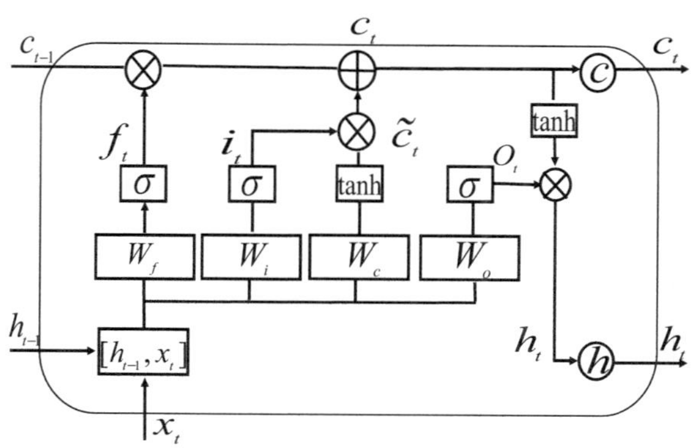

$$\tilde{C}_t = \tanh(W_c \cdot [h_{t-1}, x_t] + b_c) \tag{4}$$

$$C_t = f_t \cdot C_{t-1} + i_t \cdot \tilde{C}_t \tag{5}$$

where i_t is the output obtained by the input gate at this time; W_i and W_c are the weights of the input gate and the calculated cell state, respectively; b_i and b_c are the biases of the input gate and the calculated cell state, respectively; tanh is the hyperbolic tangent activation function.

Step 3: The output gate mainly controls the cell state input C_t and the output O_t processed by the output gate at this time, combines the two, and obtains the final output h_t, i.e.,

$$O_t = \sigma(W_o \cdot [h_{t-1}, x_t] + b_o) \tag{6}$$

$$h_t = O_t \cdot \tanh(C_t) \tag{7}$$

where W_o is the weight of the output gate at this time; b_o is the bias of the output gate at this time; and h_t is the final output of the LSTM model at this time.

2.3 *EMD-LSTM Passenger Flow Prediction Model Construction*

It is known from the analysis of the introduction that the passenger flow changes generally have a periodic pattern and also have nonlinear and non-smooth characteristics. If the original sequence is directly input, the noise of the sample itself will have an impact on the input and output of the model, reducing the accuracy of the model prediction. Therefore, it is necessary to use EMD to process the original sequence. The prediction steps of the EMD-LSTM model are shown in Fig. 2.

3 EMD-LSTM Model Prediction

3.1 *Test Data Division*

According to relevant literature studies, when the sample data volume is small (in the order of 10,000), the test sample size is generally taken as 1/6–1/3 of the total sample size in order to avoid the preference of individual samples of the model [10]. In this paper, there are 540 observations, which is a relatively small sample size, so the test sample size is chosen to be 1/6 of the total sample size, i.e., April 26–30, 2015 (90 observations) is applied as the sample test set. The subsequent outbound

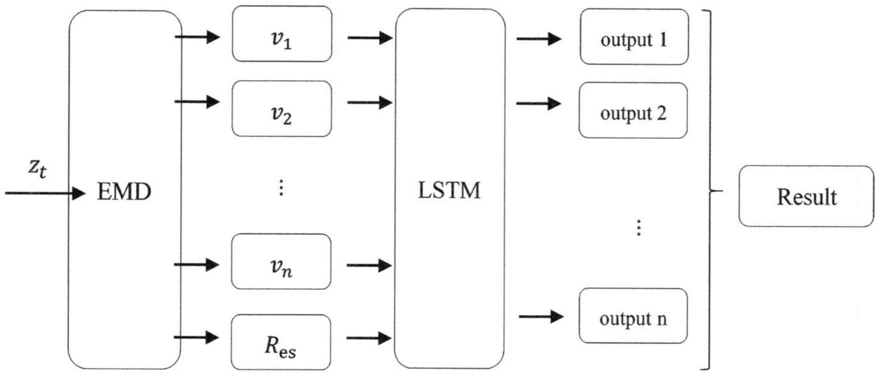

Fig. 2 Prediction steps of EMD-LSTM model

passenger flow is processed in the same way as the inbound passenger flow, and only the inbound passenger flow is used as an example here.

3.2 *Inbound Passenger Flow EMD Decomposition*

The EMD decomposition of the inbound passenger flow sequence of Xinjiekou Station is constructed at an interval of 1 h as shown in Fig. 3. The EMD decomposition has seven IMF components ($v_1, v_2, \cdots, v_7$) and one residual (R_{es}).

3.3 *LSTM Model Parameter Setting*

The LSTM model is written based on Keras, which is an efficient deep learning library. In this paper, LSTM models with multiple inputs, single output and single input, multiple outputs and multiple inputs, and multiple outputs are built with TensorFlow as the backend and Keras as the platform.

Since the data inputs of the LSTM models corresponding to each component are different, this paper illustrates the LSTM parameters taking high-frequency sequences as an example, where the number of neurons in the input layer, hidden layer, and output layer are noted as N_i, N_h, and N_O, respectively.

Input Layer

Since this paper only considers the characteristics of passenger flow variation at the time scale, the input time step at each time step is 7.

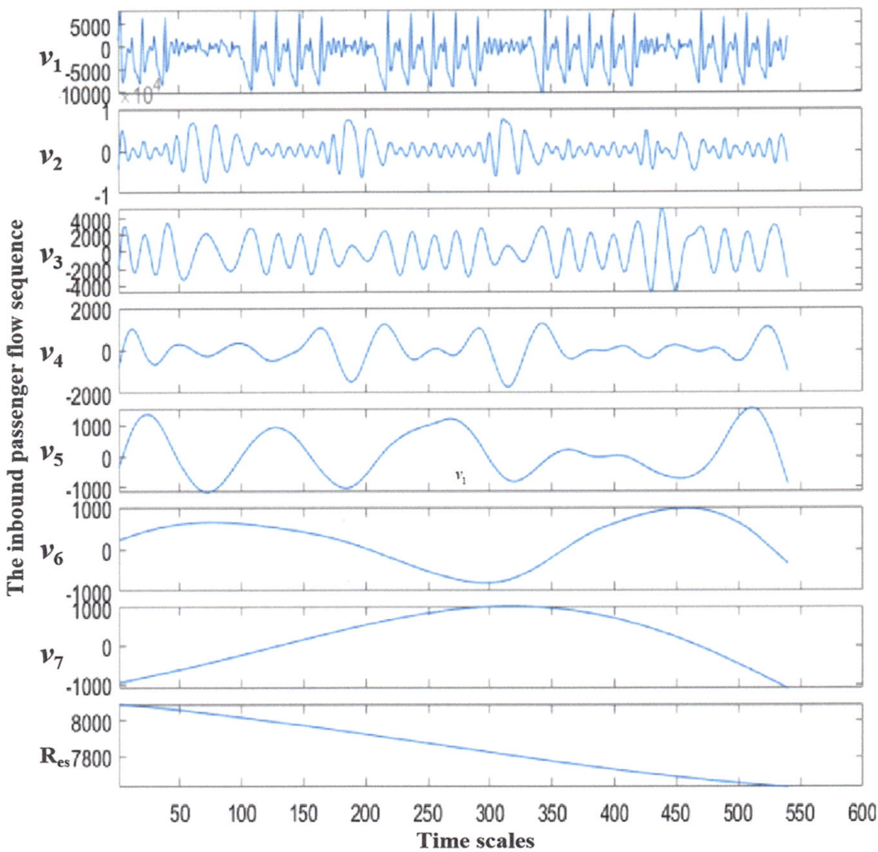

Fig. 3 EMD decomposition results of inbound passenger flow

Hidden Layers

The number of hidden layers takes the value of 3. The number of neurons in the hidden layer generally uses the empirical value:

$$N_h = \sqrt{N_i + N_o} + a \tag{8}$$

where a is a constant between 1 and 10. In this paper, different values of a are selected for testing, and finally the value of a is selected by merit.

Output Layer

The number of N_O in the output layer is mainly determined by the prediction time step. In this paper, the prediction time step is taken as 1.

3.4 Prediction Error Evaluation Index

Root-mean-square error (RMSE) and mean absolute percentage error (MAPE) are among the most commonly used indicators to evaluate the prediction accuracy, where RMSE indicates the absolute value of the difference between the observed value and the true value, which can reflect the actual situation that the prediction value deviates from the true value. The smaller the RMSE value is, the more accurate the prediction result is. MAPE indicates the degree of absolute error deviation from the true value, which reflects the merit of the model, and the smaller the MAPE value is, the more successful the model construction is. The specific formulas are shown in Eqs. (9) and (10).

$$\mathrm{RMSE} = \sqrt{\frac{1}{n}\sum_{i=1}^{n}(x_{\mathrm{pre}} - x_{\mathrm{real}})^2} \tag{9}$$

$$\mathrm{MAPE} = \frac{1}{n}\sum_{i=1}^{n}\left|\frac{x_{\mathrm{pre}} - x_{\mathrm{real}}}{x_{\mathrm{real}}}\right| \tag{10}$$

where n is the number of samples, x_{pre} is the predicted value, and x_{real} is the true value.

3.5 Comparison of Prediction Results and Error Analysis

The prediction results of inbound and outbound passenger flow of Xinjiekou Subway Station from April 26–30, 2015 obtained by applying ARIMA model, LSTM model, and EMD-LSTM model, respectively, are shown in Figs. 4 and 5. The prediction errors are shown in Table 1.

From the comparison of the prediction and error results of the three models in Figs. 4 and 5 and Table 1, it can be found that:

(1) From the time level: The peak hours are better fitted than the off-peak hours in the prediction results of the three models. This is mainly due to the stronger regularity of passenger flow changes during peak hours, so the passenger flow during peak hours is easier to predict. The prediction effect of off-peak hours is not satisfactory, which is mainly due to the fact that the vicinity of Xinjiekou Station is mainly composed of commercial and recreational areas, and the passenger flow is relatively scattered and less regular during off-peak hours.
(2) From the curve fitting effect: All three models can better reflect the volatility of the data. The ARIMA model prediction value has a certain lag, i.e., it always generates a larger error when the data fluctuates, and the error is larger near the data fluctuation. The remaining two models match the original series to a

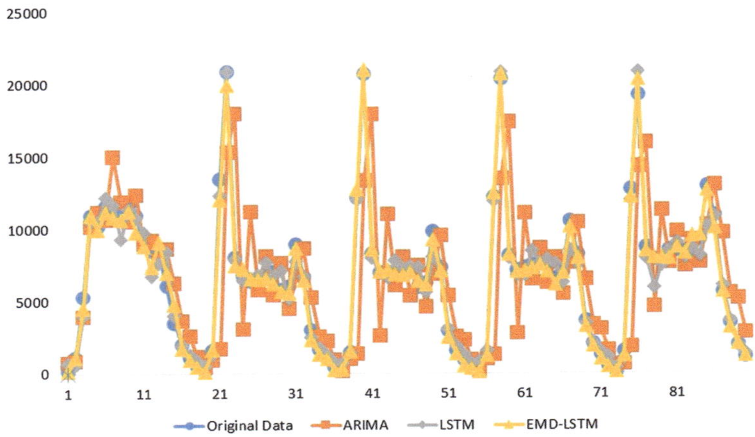

Fig. 4 Inbound passenger flow forecast results

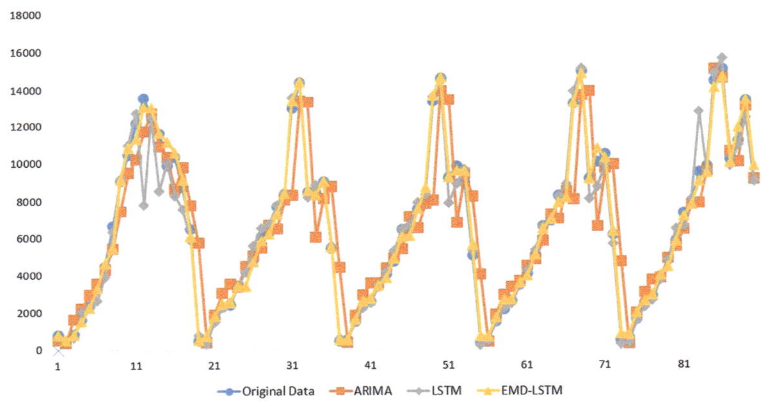

Fig. 5 Outbound passenger flow forecast results

Table 1 Error results of three models

Type of passenger flow	Models	RMSE	MAPE
Inbound	ARIMA	3823.5591	0.6701
	LSTM	860.6862	0.3135
	EMD-LSTM	592.9058	0.1669
Outbound	ARIMA	2196.4782	0.7163
	LSTM	748.7414	0.1921
	EMD-LSTM	591.3738	0.1713

higher degree, but it is difficult to distinguish which model has a more accurate prediction accuracy.

(3) From the perspective of prediction errors: Among the passenger flow prediction models, the ARIMA model has larger passenger flow prediction errors, which verifies the conclusion that the ARIMA model is more volatile in Figs. 4 and 5. The remaining two models have smaller errors of both kinds compared with the ARIMA model, which indicates that the deep learning model has stronger nonlinear fitting ability than the traditional model.

4 Conclusion

In this paper, by applying ARIMA model, LSTM model, and EMD-LSTM model to predict the passenger flow at Xinjiekou Station of Nanjing Metro through the statistics of passenger flow, the following conclusions are mainly obtained:

(1) By applying empirical mode decomposition (EMD) to decompose the original sequence, we can effectively extract local feature signals across different time scales, which significantly reduces noise interference from the original sequence. This process helps to capture the intrinsic characteristics of the data, thereby enhancing the robustness of the subsequent prediction models. In terms of prediction accuracy, the use of EMD markedly reduces the errors, leading to an improved overall accuracy of the model.

(2) Regarding the curve fitting of the prediction results, all three models demonstrate a strong ability to reflect the volatility of the data. Notably, the prediction accuracy during peak hours is superior compared to nonpeak hours. This difference is primarily due to the stronger regularity observed in peak hours, which aids in more precise modeling. The ARIMA model, however, shows some lag in its predictions, especially in response to abrupt changes, while the other two models align more closely with the original series, providing better real-time tracking of fluctuations.

(3) In terms of prediction errors, the EMD-LSTM model exhibits the lowest root-mean-square error (RMSE) and mean absolute percentage error (MAPE), indicating that it produces predictions with less dispersion and higher smoothness. This superior performance suggests that the EMD-LSTM model is particularly suitable for applications in passenger flow prediction, where both accuracy and adaptability are crucial. The feasibility of applying this model to real-world passenger flow scenarios indicates its potential for broader use in transportation planning and management.

To further enhance the accuracy and applicability of these models, future work could focus on incorporating additional influencing factors, such as weather conditions, special events, and socioeconomic indicators, into the prediction framework. These factors could provide valuable context to improve the model's ability to capture complex dynamics in passenger flow. Moreover, exploring hybrid models that combine EMD with other advanced machine learning approaches, such as attention mechanisms or graph-based neural networks, could provide deeper insights into the spatiotemporal dependencies in passenger movement. Lastly, developing adaptive models that can dynamically adjust to sudden changes, such as disruptions or anomalies, would further enhance the robustness and reliability of predictions, supporting more effective decision-making in transportation systems.

References

1. Tang, F., He, Q., Zhu, K., et al.: Study on characteristics of mass crowd evacuation in metro transfer station in a city under fire. J. Saf. Environ. 1419–1426 (2018)
2. Yang, R.: Analysis on the current situation and development strategy of urban subway construction in China. Build. Mater. Decor. 245–246 (2020)
3. Ruishan, W.A.N.G., Lantao, J.I.N.: Formation characteristics and risk management of subway stampede accidents—investigation of 15 typical accidents since. J. People's Public Secur. Univ. China (Soc. Sci. Ed.) **2016**, 141–148 (2008)
4. Wang, J., You, W., Liu, F., et al.: Fire safety evaluation of metro station based on WSR-D-S evidence theory. J. Saf. Environ. 900–906 (2018)
5. Helbing, D., Molnár, P.: Social force model for pedestrian dynamics. Phys. Rev. E. 4282–4286 (1995)
6. Daamen, W.: Modelling Passenger Flows in Public Transport Facilities. Ddlft University of Techonology, The Netherlands (2004)
7. Zhang, C.: Analysis of Passenger Flow Characteristics of Urban Rail Transit. Southwest Jiaotong University (2006)
8. Zhang, L., Chen, F.: Discussion on the operation organization of large passenger flow in subway station. Res. Urban Rail Transit 87–90 (2011)
9. Jia, S.: Research on Traffic Flow Prediction Method Based on Depth Space Time Model. Chongqing University, Chongqing (2019)
10. Pei, T., Pei, Y.: Study on Forecasting Road Traffic Volume Based on Markov Chain-BP Neural Network Model. Journal of Chongqing Jiaotong University (Natural Science Edition), 2021, pp. 35–41.

Research and Application of Mobile Phone Signaling Data in Urban Comprehensive Traffic Planning

Chunrong Xie, Tao Yang, Yanzhi Pang, and Xiali Cao

Abstract With the advent of the era of big data, the development of new urbanization continues to advance, and the functions and layout of the city itself are becoming more and more complex. Urban traffic planners are facing unprecedented challenges, and traditional traffic planning analysis techniques and methods have been difficult to cope with. How to give full play to the advantages of big data in data acquisition and analysis technology, promote the transformation of traditional traffic planning to dynamic process planning and digital planning, and realize the improvement of urban traffic planning technology and effect has become a difficult problem to be overcome in the planning field. In recent years, mobile phone signaling data has gradually emerged in the application of urban planning industry and has become one of the hot spots in planning research and application. This paper collects and preliminarily processes mobile phone signaling data through the intelligent footprint "extreme intelligence" DaaS platform and combines the urban comprehensive traffic planning system and application direction to screen and establish the application data set structure, including population analysis, urban travel, intercity travel, hub passenger flow analysis, scenic spot or other special place passenger flow analysis, traffic investigation and analysis of six categories and 14 subcategories, one by one given the "letter+number" identification code. ArcGIS is used to calculate and analyze various data sets and output map visualization. A set of application guidelines for mobile phone signaling data in traffic planning is collected and sorted out; the relevant data result standards are unified, and the processing, analysis, and application of mobile phone signaling data are standardized. It provides guidance and reference for traffic planning management-related technology research and planning application.

Keywords Traffic planning · Mobile phone signaling data · Application data sets · Application guidelines

C. Xie · T. Yang (✉)
College of Traffic and Transportation, Nanning University, Nanning, China
e-mail: 734348102@qq.com

Y. Pang · X. Cao
Guangxi Key Laboratory for International Join for China-ASEAN Comprehensive Transportation, Nanning University, Nanning, China

K. Subramaniyam and M. Paprzycki (eds.), *Smart Cities and the Internet of Things*, Internet of Things, https://doi.org/10.1007/978-3-032-08290-9_2

1 Introduction

Since 2016, the Global Big Data Summit, China Big Data Application Conference, and China International Big Data Conference have been held in succession, which means that big data has received extensive attention. With the advent of the era of big data, the development of new urbanization continues to advance, and the functions and layout of the city itself are becoming more and more complex. Urban traffic planners are facing unprecedented challenges, and traditional traffic planning analysis techniques and methods have been difficult to cope with. How to give full play to the advantages of big data in data acquisition and analysis technology, promote the transformation of traditional traffic planning to dynamic process planning and digital planning, and realize the improvement of urban traffic planning technology and effect has become a difficult problem to be overcome in the planning field. Mobile phone signaling data has various advantages such as wide network coverage, stable and reliable data, large sample size, high spatial resolution, and strong dynamics [1]. Cities are places with concentrated population and mobile phone users. Compared with other big data, mobile phone signaling has a better fit with the field of urban planning, which is reflected in the full sample, real-time, dynamic, and no group differences. Therefore, in recent years, mobile phone signaling data has gradually emerged in the application of urban planning industry and has become one of the hot spots in planning research and application.

At present, many studies have explored the application of mobile phone signaling data in bus station optimization [2], travel origin–destination (OD) matrix analysis [3], passenger flow identification and prediction [4], analysis of residents' travel time and space characteristics [5], analysis of urban occupation and residence and urban spatial characteristics [6], and analysis of commuting changes during special periods. However, these studies mainly focus on a certain subdivision of urban planning or traffic planning or a certain characteristic index, which is difficult to systematically reflect the application of mobile phone signaling data in traffic planning. Some scholars have studied the macro planning and application direction of a city by using mobile phone signaling data [7]. However, these studies usually focus on macro planning or application directional exploration and have not formed a systematic data application structure and guidelines.

In summary, the current research on the application of mobile phone signaling data focuses on a certain subdivision of urban macro planning or traffic planning. There are few related studies that systematically integrate traffic planning and mobile phone signaling data to form application guidelines. Therefore, this paper collects and preliminarily processes mobile phone signaling data through the intelligent footprint "extreme intelligence" DaaS platform, combines the urban comprehensive traffic planning system and application direction to screen and establish the application data set structure, and uses ArcGIS to perform various data sets. The operation analysis and map visualization output are carried out, and a set of structured and standardized application guidelines for mobile phone signaling data in traffic planning

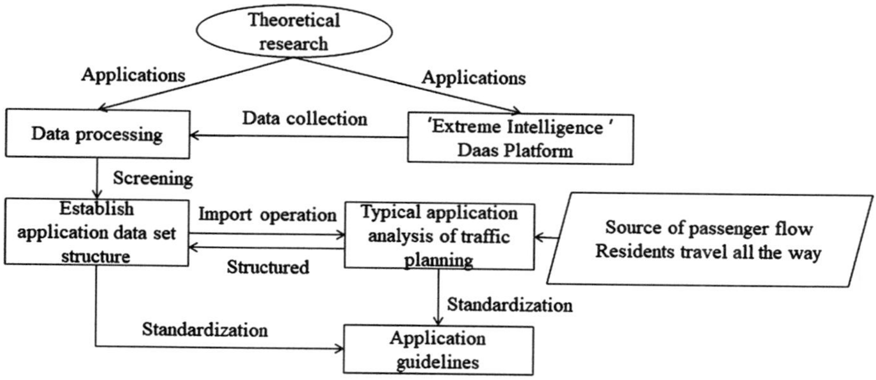

Fig. 1 Research framework

is collected and sorted out, which provides normative technical application guidance for the processing, analysis, and application of mobile phone signaling data, offering guidance and reference for traffic planning management related technology research and planning application. This promotes the transformation of traditional traffic planning into dynamic process planning and digital planning. The research framework of this paper is shown in Fig. 1.

2 Theoretical Basis

Mobile phone signaling data, as a kind of spatiotemporal big data related to human activities, has the characteristics of continuous time, complete space, passive collection, and real non-perception [8]. The mobile phone signaling data mainly comes from the mobile communication system of the three major communication operators. There are many mobile signal base stations in the city. Each base station records the mobile phone user information that appears in its signal range. The information received by the base station within a certain range is aggregated to identify the activity trajectory of a mobile phone user in this range. There are several types of mobile phone signaling data, such as call detail data, including user attributes such as attribution, age, etc. This data is generally used in research and applications mainly to identify the user's spatial location [9]. Compared with other types of big data sources, mobile phone signaling data has outstanding advantages in the analysis of urban residents' activity characteristics, hub passenger flow distribution, population migration and distribution, traffic planning, and decision-making.

According to the data processing flow of mobile phone signaling data, in order to realize the fusion application of signaling data and traffic planning, a series of technical methods and algorithms, such as Thiessen polygon mapping method, feature recognition algorithm, time series analysis method, and cluster analysis method, need to be applied to carry out a lot of data processing work, so as to obtain an effective application data set that can be used for industry application analysis. This study uses the Thiessen polygon approximation to obtain the spatial coverage of each base station and then maps the base station to the traffic zone and the road traffic network [10]. It identifies the spatial and temporal location of each mobile phone user in the road traffic network and then performs cluster analysis and travel chain processing to continuously and completely reproduce the trajectory of each user in a certain period of time. Hadoop distributed cluster is used to construct a massive data feature recognition and extraction algorithm to identify the residence behavior of mobile phone users within a certain range and to screen out the user's place of residence, leisure place, or interest area, so as to provide data support for the next resident travel OD analysis.

3 Data Acquisition and Processing

3.1 Data Acquisition

The data acquisition in this paper is an effective link to collect the original mobile phone signaling data by using the intelligent footprint "extreme intelligence" DaaS (data integration) platform. These effective links have the attributes of service base station number, switching time, location, and so on. According to the requirements of time period, range, and statistical granularity, the basic data set in the study period is counted. Then, according to the application requirements of the industry, the data modeling operation is carried out to form the application data set, and the ArcGIS software is used to import the data for secondary analysis and visualization.

3.2 Data Processing

The mobile signaling data processing flow of this study is as follows: data cleaning, base station cell positioning, geospatial location mapping, travel chain identification, and partition statistics and result expansion [11].

(1) Data cleaning: First, the scope of the study area is delineated, the data in the latitude and longitude range of the study area is screened, and the data that is not within the scope of the study area is filtered. Then, the unreasonable data such as duplicate data, ping-pong switching data, missing data, static redundant data, drift data, and so on are filtered and optimized [12].
(2) Base station cell positioning: The location of base station cells should be selected in the main urban area of the city with high population density and high base station density. Therefore, the base station service density of this study is 500 m, that is, the statistical space granularity is 500 m * 500 m grid [13, 14].

(3) Geospatial location mapping of individual mobile phone time series: The mobile phone base station signal is mapped to the urban geospatial space to obtain the movement of mobile phone users in the urban geospatial space. The Thiessen polygon approximation method is generally used to obtain the spatial coverage of the base station and map the traffic zone to the base station [15].

(4) User travel chain identification: According to the time, the user signaling data is extracted, and then the all-weather movement trajectory of each user's mobile phone is obtained. Then the travel chain recognition model can be constructed [16]. In this model, the mobile phone movement trajectory can be divided into several continuous travel points, and then the starting point and stay point of each user's travel can be identified.

(5) Partition statistics and result expansion: First, the spatial analysis unit is divided on demand, that is, the traffic partition, and then the corresponding relationship between the base station cell and the traffic partition is built. According to the traffic partition requirements, the user travel records are summarized and counted accordingly, and then the travel application data set based on the traffic partition is obtained. In order to ensure the validity of the mobile phone signaling sample, it is required to implement the sample expansion processing and verify the accuracy of the sample expansion results.

4 Research on the Structure of Application Data Set Based on Traffic Planning

4.1 Research on Application Data Set Structure

Based on the above data acquisition and processing, various application data sets can be output, and the application data set structure can be established in combination with the "Standard for Urban Comprehensive Transport System Planning (GB/T 51,328-2018)" [17], including population analysis, urban travel, intercity travel, hub passenger flow analysis, scenic spots or other special place passenger flow analysis, and traffic survey analysis of six categories and 14 subcategories, and the "letter and number" identification code is given one by one, as shown in Fig. 2. The corresponding data sets and data attributes of each type of application are sorted out to form application technical guidelines, which are convenient for urban traffic planners and managers to refer to.

4.2 Typical Application Data Set Export Analysis

The population spatial distribution (A1) data set export uses the travel chain analysis model to screen the travel characteristics of different travel purposes of the mobile phone user group and to identify the travel characteristics of the resident population

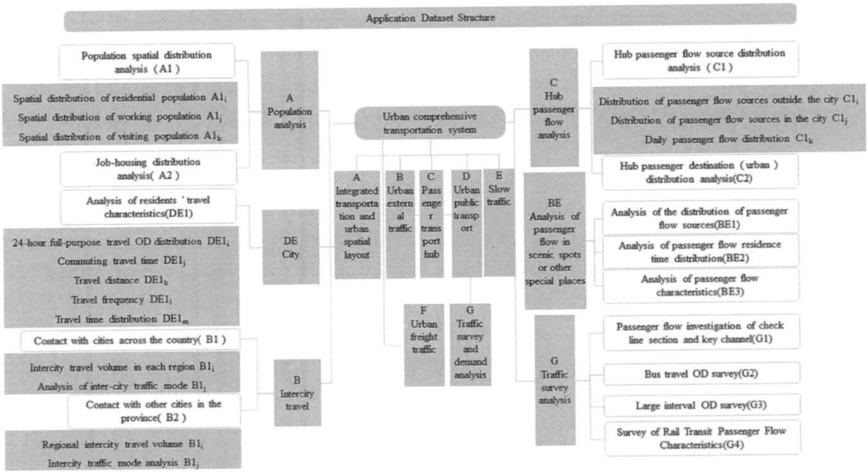

Fig. 2 Application data set framework diagram

within the model's setting parameters. The user's work-and-rest behavior characteristics and residence time can be calculated and analyzed by using the user's stay model. Through the screening of multiple sets of data, the user's permanent residence and workplace are identified, and then the screening set value of this feature is derived. Finally, the distribution of the permanent population's residence and workplace in each traffic district and the job-housing ratio are obtained. After anonymous extraction, high automation, and deep noise reduction processing, the mobile phone signaling data is processed into grid data reflecting the spatial distribution and connection of users' jobs and residences as the basic data of this paper, as shown in Table 1.

The resident travel feature (DE2) data set export is a spatial aggregation of the user's travel trajectory at a distance of 500 m. The point where the residence time

Table 1 List of residential/working population spatial distribution data set (30 days)

Date	Traffic zone number	Original statistics of the resident population	Statistical value of residential population after sample expansion
20,230,101	1	39/32	216/177
20,230,101	10	700/265	3871/1465
20,230,101	100	4610/1104	25,493/6105
20,230,101	102	907/375	5016/2074
20,230,101	104	2095/490	11,585/2710
20,230,101	106	2554/711	14,124/3932
……	……	……	……

exceeds 30 min is used as a stay activity point, and the user's one-day travel chain trajectory and travel times and travel time can be obtained. The intercity travel ($B1$) data set is derived by identifying the user's permanent residence during the study period, counting, and correcting the error to obtain the number of users living in Nanning. Through the comprehensive analysis of multidimensional factors such as user ownership and origin, the hometown of the resident population is identified. The travel of O and D points in the total travel OD in different cities is marked as intercity travel, and the travel volume from Nanning to the whole country and from the whole country to Nanning is obtained. The export of hub passenger flow ($C1$) data set is to delineate the monitoring area of the hub according to the coverage of the base station signal and the distribution of the functional areas of the transportation hub [18]. The residence time threshold is set, and the data on the number of hubs, the change rate, and the passenger flow of other provinces during this period is analyzed and exported. Based on the analysis of resident population and transit personnel, the hub passenger flow data is further processed. The relevant passenger flow monitoring indicators are eliminated from the resident population and transit population to obtain the actual hub passenger flow.

5 Mobile Phone Signaling Data and Traffic Planning Integration Application Example

Based on the above research, the Nanning's distribution of all-way travel of urban residents is taken as an example for application analysis. Then, based on industry standards and applications, the application data and attributes of various application data sets in the field of traffic planning are collected, including statistical space granularity, statistical time granularity, data statistical caliber, related index calculation model, data description, data field description table, and related drawing examples. Therefore, the application guidelines of mobile phone signaling data in the field of traffic planning are formed as an application supplement of national industry standards [19].

5.1 Application Data Set Import

Through data processing and analysis, a 24-h full-purpose travel data set is obtained, as shown in Table 2. This data set is imported into ArcGIS for operational analysis and processing.

Table 2 24-h full-purpose travel distribution table

Date	Travel time	Traffic zone number of departure place	Traffic district number of arrival place	Number of trips
20,230,113	5	1	41	6
20,230,113	9	1	214	6
20,230,113	13	1	531	6
20,230,113	16	1	123	6
20,230,113	17	1	443	6
20,230,113	20	1	59	6
20,230,114	11	1	202	6
……				

5.2 Application Analysis

Traditional methods usually use land-use status data to analyze the implementation of spatial layout, but they cannot judge the spatial relationship between areas and groups, mainly based on qualitative judgment center system [20]. Through the mobile phone signaling data, the residents' travel links and intensity between the various areas of the city can be measured, so as to quantitatively analyze the urban spatial links. Nanning plans to form five comprehensive service centers (Langdong, Wuxiang, Chaoyang, Wuxu, and Nanbu) and two new town centers (Wuming and Liujing). By analyzing the mobile phone signaling data from January to March 2023, it can be seen that the residents' travel in Nanning is mainly concentrated in the old urban area and between the old urban area and the surrounding new area. Among them, the travel volume between Fengling and the central group is the largest, followed by the travel between the central group and the Jiangnan and Chengbei groups, and the travel between the Fengling and Jiangnan groups, forming the three centers of the central group (Chaoyang), Fengling group (Langdong), and Tinghong group. Wuming District and Jiangnan Tinghong District, Wuxiang New District and Fengling District, Yudong District, Sansitang District, and Fengling District, as well as the central group, show a high intensity of contact. Wuming, Wuxiang, Sansitang, and other new district centers are gradually formed, but the overall radiation capacity is still weak, as shown in Fig. 3. There is still a certain gap between the current urban center system and the future planned center system.

Based on the above analysis, it can be seen that the analysis and application of mobile phone signaling data can effectively solve the problems that are difficult to solve in traditional planning. For example, it is difficult to quantitatively analyze and judge the main traffic corridor and urban center system. Traffic survey has long periodicity, heavy workload, small sample size, high cost, and insufficient granularity and dimension. Through the mobile phone signaling data, the traffic indicators such as residents' travel connection and intensity, hub passenger flow, and intercity travel

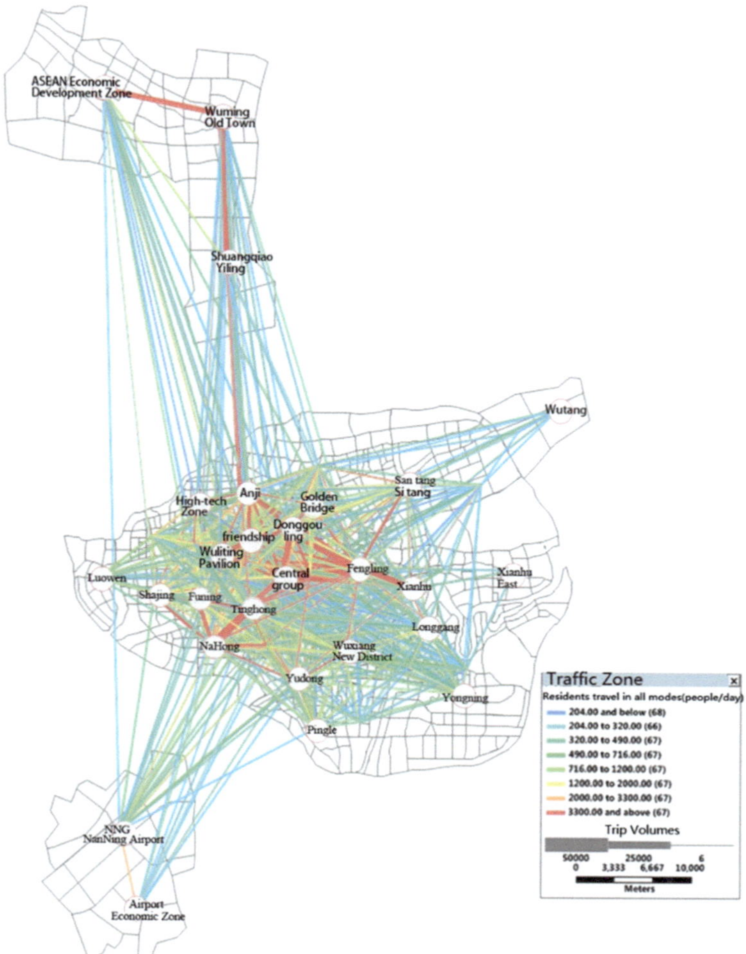

Fig. 3 Distribution map of daily average residents' all-mode travel in Nanning city

volume between various areas of the city can be measured. The measurement results have a high degree of fitting and accuracy with the actual situation.

6 Conclusion

In this paper, based on the intelligent footprint "extreme intelligence" DaaS platform, the whole-process processing, analysis, and application of mobile phone signaling data are carried out through examples, from basic theory to practical application analysis, and then from practice to summarizing the application rules and standards, finally forming a standardized guidance document.

In this paper, combined with the application analysis of examples, it can be seen that the calculation results of mobile phone signaling data have a high degree of fitting and accuracy with the actual situation and can effectively solve the problems that are difficult to solve in traditional planning. In particular, judging the transformation of many aspects such as traffic main corridors and urban center systems from traditional qualitative analysis to "quantitative and qualitative" analysis makes the analysis conclusions more comprehensive and scientific. In a word, the application cases fully verify the feasibility of applying mobile phone signaling data technology to traffic planning. The traffic planning application data set structure and technical guidelines built in this paper are essential for standardizing and guiding the application of traffic planning data. The data set structure and technical guidelines are relatively high for large or super-large cities with large population density and medium or above economic level. The guidance is relatively high, because this type of urban residents has a high proportion of smart phones and a high coverage of communication base station construction. Overall, this application guideline has good targeted guidance and scientific rationality. It can provide guidance and reference for urban traffic planning, design, and management and can be used as one of the application supplements to national industry standards. With improvements in the coverage and data quality of signaling data, its analysis accuracy will continue to improve, and better application results can be achieved. In the future, signaling data will have a wider application prospect in traffic planning.

Acknowledgements This research is supported by two projects: (1) Guangxi Science and Technology Base and Talent Project for Guangxi Science and Technology Plan Project: Construction of Guangxi Transportation New Technology Transfer Center Platform (No. Guike AD23026029); (2) Construction of Scientific and Technological Innovation Bases for Science Research and Technology Development Plan Project of Nanning City: Intelligent Operation and Maintenance of Transportation Infrastructure and Health Monitoring Platform (No. 20223230).

References

1. Sudani implementation and application effectiveness of Hunan real-time transportation and tourism processing platform based on mobile signaling data. Transp. World (29), 21–24 (2020). https://doi.org/10.16248/j.cnki.11-3723/u.2020.29.007
2. Ge, H., Lv, Y., Jiao, P.: Optimization method for commuter bus stops in medium-sized cities based on signaling data. Traffic Inf. Secur. **42**(01), 142–149 (2024)
3. Jiang, Y., Lei, M., Gu, M., et al.: Typical daily OD matrix analysis based on mobile signaling data. Compr. Transp. **46**(06), 123–129 (2024) https://doi.org/10.20164/j.cnki.cn11-1197/u.2024.06.018
4. Chen, Y., Zhang, Y., Zhang, Y., et al.: A framework for identifying and predicting air rail intermodal passenger flow in urban agglomerations based on signaling data. J. Beijing Jiaotong Univ. 1–12 (2024)

5. Zhao, P., Liu, W., Yu, L., et al.: Research on the spatiotemporal characteristics of small city residents' travel based on mobile signaling data: a case study of Tongren City. West. J. Hum. Settl. Environ. **39**(03), 155–161 (2024). https://doi.org/10.13791/j.cnki.hsfwest.20240322
6. Liu, L., Yan, C., Chen, Z., et al.: The application of mobile signaling data in urban and transportation planning in Changsha city. In: Academic Committee on Urban Transportation Planning of China Urban Planning Society Innovation driven and intelligent development—Proceedings of the 2018 China Urban Transportation Planning Annual Conference Changsha Planning, Surveying, Design and Research Institute, vol. 10 (2018)
7. Xu, Z.: Application and reflection on mobile signaling data in transportation planning. Electron. Compon. Inf. Technol. **4**(08), 75–77 (2020). https://doi.org/10.19772/j.cnki.2096-4455
8. Liu, Y.: Exploration of smart city application scenarios based on mobile signaling. China Inf. Ind. (05), 122–124 (2024)
9. Dong, J.: Design and Implementation of a Regional Human Flow Monitoring Platform Based on Big Data. Beijing Jiaotong University (2018)
10. Fu, L.: Analysis of resident travel survey based on mobile signaling data: a case study of shijiazhuang. Transp. Commun. (Acad. Ed.) (02), 68–71 (2018)
11. Tang, X., Zhou, T., Lu, B.: Analysis of large scale human flow movement based on mobile phone signaling. J. Chongqing Jiaotong Univ. (Nat. Sci. Ed.) **36**(01), 82–87+109 (2017)
12. Huang, Y., Gong, W.: Travel behavior analysis based on mobile phone signaling data of urban public transportation. People's Public Transp. (15), 89–92 (2024)
13. Li, Z., Lei, M., Ji, X., et al.: Urban traffic model and application based on mobile signaling data. Compr. Transp. **46**(10), 182–188 (2024)
14. Zhang, K., Tang, J., Duan, Z., et al.: Research on the application of mobile signaling big data in macro traffic models. Transp. Commun. **36**(S1), 226–231+241 (2023)
15. Wang, D., Xu, W., Cai, Z., et al.: Bayesian optimization of travel chain recognition based on mobile signaling data. J. Zhejiang Univ. (Eng. Ed.) **58**(11), 2393–2405 (2024)
16. Chang, Q.: The current situation and analysis of using mobile signaling big data for traffic investigation. Theor. Res. Urban Constr. (Electron. Version) (29), 59 (2019)
17. GB/T 51328-2018. Standard for urban comprehensive transport system planning. Minist. Hous. Urban-Rural. Dev. (2018)
18. Zhong, J.: Research on Integrated Traffic Hub Passenger Flow Monitoring Technology Based on Mobile Signaling. Chongqing Jiaotong University (2017)
19. JT/T 1182.2-2018. Data collection and exchange service for road network operation monitoring based on mobile signaling-Part 2: data collection. Minist. Transp. (2018)
20. Huang, X., Li, Y., Liu, Z., et al.: Refined identification of occupational residential areas based on residents' travel characteristics. Netw. Secur. Data Gov. **43**(08), 44–48 (2024)

Prediction of Taxi-Out Time Based on LSTM-XGBoost

Guoxin Zhang, Guan Lian, Benxiao Lou, and Wenyong Li

Abstract With the rapid development of the global aviation industry, reducing flight delays has become a critical challenge. Aircraft taxi-out time, which directly impacts the utilization of runways and taxiways, is one of the most significant factors influencing the operational efficiency of airport field operations. This study takes Beijing Capital International Airport as a case study, identifying model characteristic parameters through a comprehensive analysis of historical departure data. An integrated long short-term memory–extreme gradient boosting (LSTM-XGBoost) model is proposed to predict taxi-out times, with further optimization achieved using the sparrow search algorithm (SSA). To evaluate the model's generalizability, historical data from Zhengzhou Xinzheng International Airport are utilized for validation. Comparative analysis of prediction evaluation metrics reveals that the integrated LSTM-XGBoost model outperforms standalone LSTM and XGBoost models, demonstrating superior prediction accuracy and robustness.

Keywords Airport operations · Taxi-out time · Integrated model

G. Zhang · G. Lian (✉)
School of Architecture and Transportation Engineering, Guilin University of Electronic Technology, Guilin, China
e-mail: lianguan@guet.edu.cn

G. Zhang
e-mail: zhangguoxin@mails.guet.edu.cn

B. Lou
School of Traffic and Transportation Engineering, Central South University, Changsha, China
e-mail: benxiaoLou@csu.edu.cn

W. Li
Guangxi Vocational and Technical College of Communications, Nanning, China
e-mail: Traffic@guet.edu.cn

K. Subramaniyam and M. Paprzycki (eds.), *Smart Cities and the Internet of Things*, Internet of Things, https://doi.org/10.1007/978-3-032-08290-9_3

1 Introduction

With the rapid advancement of the global air transport industry in recent years, the reduction of flight delays has become a pressing concern. The taxi-out time of an aircraft is directly linked to the utilization of runways and taxiways, exerting a substantial influence on the overall operational efficiency of airports. Consequently, the ability to predict aircraft taxi-out times with scientific precision holds considerable importance for minimizing airport operational costs and enhancing the effectiveness of airport operations.

For the analysis of taxi-out time prediction and influencing factors, Yin et al. [1] classified the existing mainstream models and methods, analyzed the main ideas and scope of application of various methods, and proposed the future development direction of departure glide time prediction. Qiao et al. [2] proposed a skidding time prediction method based on adaptive scene matching rules and formed a feature set of skidding time influencing factors consisting of nine key features. Balakrishna, Lim et al. [3] used gradient-boosted decision tree (GBDT) model to predict flight departure taxi-out times and used SHAP values to analyze and interpret the model. Floris Herrema et al. [4] focused on the use of neural networks, regression trees, reinforcement learning, and multilayer perceptron methods to predict taxi-out times. Lian et al. [5] proposed two improved support vector regression (SVR) methods based on swarm intelligence algorithm optimization. Results show that the SVR algorithm based on IFA optimization has a better modeling effect and better accuracy when dealing with abnormal taxiing time situations. Chen et al. [6] proposed a prediction algorithm based on a fuzzy rule system and constructed a nonlinear regression model to predict taxiing time. Kim et al. [7] predicted aircraft departure taxi-out times using a random forest algorithm. Okwir et al. [8] applied the SHAP method to characterize the importance of all independent variables and verified that linear regression and elastic networks are the most effective machine learning models for achieving high prediction accuracy within the CDM framework. Tang et al. [9] proposed a surface traffic flow computation method considering arriving and departing flights using gradient-boosted regression trees to predict glide times and Lasso to analyze feature importance. Yang et al. [10] proposed a multitask dynamic spatiotemporal graphical attention network (MT-DSTGAN) that incorporates future airport traffic flows to improve taxi-out time prediction accuracy. Park et al. [11] gave in-depth consideration to airport weather as an influencing factor, including wind direction, customs, visibility, etc., for taxi-out time prediction.

A comprehensive analysis of international research results shows that the prediction of aircraft departure taxi-out time based on machine learning tends to use a single model, which has a bottleneck in improving the accuracy of taxi-out time prediction. Therefore, this study proposes an integrated learning model of long short-term memory–extreme gradient boosting (LSTM-XGBoost) based on scene data mining and analysis and uses the sparrow search algorithm (SSA) to optimize its parameter combination, which achieves better prediction accuracy than a single model.

2 Characteristics and Evaluation Indicators

Due to the differences in runway layout and dispatch methods at different airports, the impact of field conditions must also be considered when analyzing the characteristics of departure taxi-out time.

2.1 Data

The airport selected in this paper is Beijing Capital International Airport, which is the highest-level civil airport in China and also an important large hub airport with an annual passenger throughput of over 50 million. Capital International Airport has three runways, namely 01/19, 36R/18L, and 36L/18R. Among them, 36R/18L is the main runway for takeoffs, and this runway is selected as the research object in the subsequent analysis.

The data used in this study consist of operational data from the departure scenes at Beijing Capital International Airport on November 13, 14, and 15, totaling 1471 records. These dates were all weekdays, and mid-November did not coincide with peak passenger flow periods such as holidays, making the data more broadly applicable.

At the same time, in order to avoid the impact of too sparse data on the prediction accuracy of the model, this study only uses the departure data from 6:00 to 22:00 every day for analysis. The taxiing time in the text refers to the time interval from the actual removal of the gear to the time when the wheels leave the ground during taxiing and takeoff. The formula is as follows:

$$T_{\mathrm{st}} = T_{\mathrm{at}} - T_{\mathrm{rp}} \tag{1}$$

where T_{st} indicates taxi-out time, T_{at} indicates actual takeoff time, and T_{rp} indicates real push time.

2.2 Feature

Through the analysis of scene data, the model characteristics selected in this study are the number of aircraft in the land (LandNumber), the length of the aircraft queue (Queuelength), the delay time (Delay), the real push time (Real Push Time), and the planned takeoff time (Plan Takeoff Time). Some of the data for the characteristics are listed in Table 1 for subsequent prediction.

The actual launch time and scheduled departure time are shown in the table. To improve the accuracy of the forecast and facilitate analysis, the 24-h data is converted

Table 1 Feature statistical data

Feature	Data type		
	Median	Average	Maximum
Land number	6	6.96	43
Queue length	3	5.16	33
Delay	17	21	239
Real push time	446	459	963
Plan takeoff time	449	458	960

to the minute system and then uniformly subtracted by 360 min, which ensures that the data is counted from 6:00 every day.

2.3 Index

In order to scientifically and comprehensively evaluate the prediction results, this study selected four evaluation indicators: absolute error percentage accuracy, mean absolute error (MAE), mean squared error (MSE), root-mean-square error (RMSE), and determination coefficient (R^2).

3 Methods

Integrated learning is a learning method that combines multiple weak learners to construct a combined model through strategies such as sample weighting and learner weighting. It requires individuals to have a certain degree of accuracy and variability. Studies have shown that among many learning tasks, the performance of ensemble learning algorithms significantly improved compared to individual learners such as decision trees and support vector machines. Therefore, this study constructs a taxi-out time prediction model based on long short-term memory networks (LSTM) and extreme gradient boosting (XGBoost), two learners with good regression prediction results, and uses the sparrow search algorithm (SSA) to optimize the ensemble model. Among them, the sparrow search algorithm is a swarm intelligence algorithm inspired by the foraging behavior of sparrows, which is optimized for the hyperparameter tuning of the model and can further improve the prediction accuracy of the model.

3.1 LSTM Regression

LSTM, or long short-term memory, is a neural network model capable of capturing long-term dependencies in continuous data. The model is an improvement of the traditional recurrent neural network (RNN), and its structure can improve the problems of excessive weight influence and gradient disappearance during the training of RNN, with a better ability to deal with long time sequences. Compared with RNN, LSTM adds three gates, namely forgetting gate, input gate, and output gate. LSTM unit can keep valid data for a long time, capture long-term dependencies better than traditional RNN, and exhibit better time sensitivity in regression prediction than traditional learners.

Through the first step of the forgetting gate, it is decided to selectively forget information using an activation function by inputting the discarded information from the existing state. The expression is as follows:

$$f_t = \sigma\left(W_f \bullet \left[h_{t-1}, x_t\right] + b_f\right) \tag{2}$$

where σ represents the Sigmoid activation function, W_f is the weight parameter of the forget gate connecting the input and the hidden layer neuron, and b_f is the bias of the hidden layer neuron of the forget gate.

The input gate mainly determines which information is taken and updated, and its expression is shown in (3) and (4).

$$i_t = \sigma\left(W_i \bullet \left[h_{t-1}, x_t\right] + b_i\right) \tag{3}$$

$$\widetilde{C}_t = \tanh\left(W_C \bullet \left[h_{t-1}, x_t\right] + b_C\right) \tag{4}$$

where tanh is a type of activation function that limits the output to [−1,1], W_i and W_C are parameters that connect the input to the neurons in the hidden layer, and b_i and b_C are the biases, respectively.

The output gate mainly consists of a Sigmoid function determining the output part, which is then multiplied by the updated cell state to obtain the final output.

3.2 XGBoost Regression

XGBoost is an efficient gradient boosting tree algorithm commonly used to solve regression and classification problems. It combines gradient boosting and regularization techniques to improve model performance and reduce the risk of overfitting. The idea of the algorithm is to generate a new tree at each iteration to fit the residual of the previous tree and to make the predicted value close to the true value through continuous iteration.

In regression prediction, the model uses parallel and distributed computing optimization algorithms to effectively reduce training time. At the same time, through built-in processing methods, the model does not need to pre-fill missing values and can directly achieve better prediction results when the data is incomplete. The objective function can be expressed as:

$$L(\theta) = \sum_{i=1}^{n} l(y_i, \hat{y}_i) + \sum_{k=1}^{K} \Omega(f_k) \tag{5}$$

where $l(y_i, \hat{y}_i)$ is the loss function, which measures the difference between the true value y_i and the predicted value $\hat{y}_i$, and $\Omega(f_k)$ is the regularization term.

3.3 LSTM-XGBoost Regression

In this study, based on the principle of integrated learning, the LSTM-XGBoost integrated model is proposed by combining the error weight coefficients of two base learners, LSTM and XGBoost, which is implemented as follows:

Initially, the preprocessed dataset is partitioned into three subsets: training, validation, and testing. Subsequently, the LSTM and XGBoost models are trained on the training subset to generate the models f_{LSTM} and f_{XGB}, respectively. To enhance the prediction accuracy, the sparrow search algorithm (SSA) is employed to optimize the hyperparameters of each model based on the validation set. This optimization process is conducted using a specified number of iterations (iter) and a defined population size for the algorithm. Through this iterative optimization, the optimal parameter configurations for both models are identified, and the respective errors on the validation set are recorded for each model. This approach allows for fine-tuning of the models to improve their predictive performance before testing.

If the true value of the validation set is set to y_v, and the model predictions are y_v^{LSTM} and y_v^{XGB}, then the errors of the model on the validation set are shown in Eqs. (6) and (7):

$$\mathrm{MAE}_v^{\mathrm{LSTM}} = \left| y_v^{\mathrm{LSTM}} - y_v \right| \tag{6}$$

$$\mathrm{MAE}_v^{\mathrm{XGB}} = \left| y_v^{\mathrm{XGB}} - y_v \right| \tag{7}$$

The weight of the combination is determined by the error of the two models on the validation set:

$$W^{\mathrm{LSTM}} = \frac{\mathrm{MAE}_V^{\mathrm{XGB}}}{\mathrm{MAE}_V^{\mathrm{LSTM}} + \mathrm{MAE}_V^{\mathrm{XGB}}} \tag{8}$$

$$W^{\text{XGB}} = \frac{\text{MAE}_V^{\text{LSTM}}}{\text{MAE}_V^{\text{XGB}} + \text{MAE}_V^{\text{LSTM}}} \tag{9}$$

If the prediction accuracies of the two algorithms on the validation set exhibit a substantial disparity, the combined predictive performance of the two algorithms is expected to approximate the median of their individual accuracies. To further enhance prediction accuracy, the algorithm yielding the superior result can be selected as the final model. Consequently, this study introduces a threshold θ to determine the appropriateness of combining the models. Specifically, if the following condition (10) is met indicating that the relative difference in prediction accuracies between the two algorithms exceeds the threshold θ, the combination process will not be undertaken.

$$\frac{\left|\text{MAE}_v^{\text{LSTM}} - \text{MAE}_v^{\text{XGB}}\right|}{\max\left(\text{MAE}_v^{\text{LSTM}}, \text{MAE}_v^{\text{XGB}}\right)} > \theta \tag{10}$$

$$\text{MAE}_v^{\text{LSTM}} > \text{MAE}_v^{\text{XGB}},\ W^{\text{LSTM}} = 0,\ W^{\text{XGB}} = 1$$

$$\text{MAE}_v^{\text{LSTM}} < \text{MAE}_v^{\text{XGB}},\ W^{\text{LSTM}} = 1,\ W^{\text{XGB}} = 0$$

After the prediction model is tested on the test set, the predicted values on the test set are y_T^{LSTM} and y_T^{XGB}, respectively. Combining the definitions above, the expression for the prediction result y_T^{EM} obtained after model integration is as follows:

$$y_T^{\text{Em}} = W^{\text{LSTM}} y_T^{\text{LSTM}} + W^{\text{XGB}} y_T^{\text{XGB}} \tag{11}$$

In summary, the flowchart of the LSTM-XGBoost model can be drawn as shown in Fig. 1.

4 Results

4.1 Feature Importance

In this study, a random forest model was employed to evaluate the significance of input features during the model training process. This approach enables a visual representation of the importance of each feature, thereby supporting informed decisions for subsequent model adjustments and optimization. The calculated feature importances are presented in Table 2, providing insight into the relative influence of each feature on the model's predictive performance.

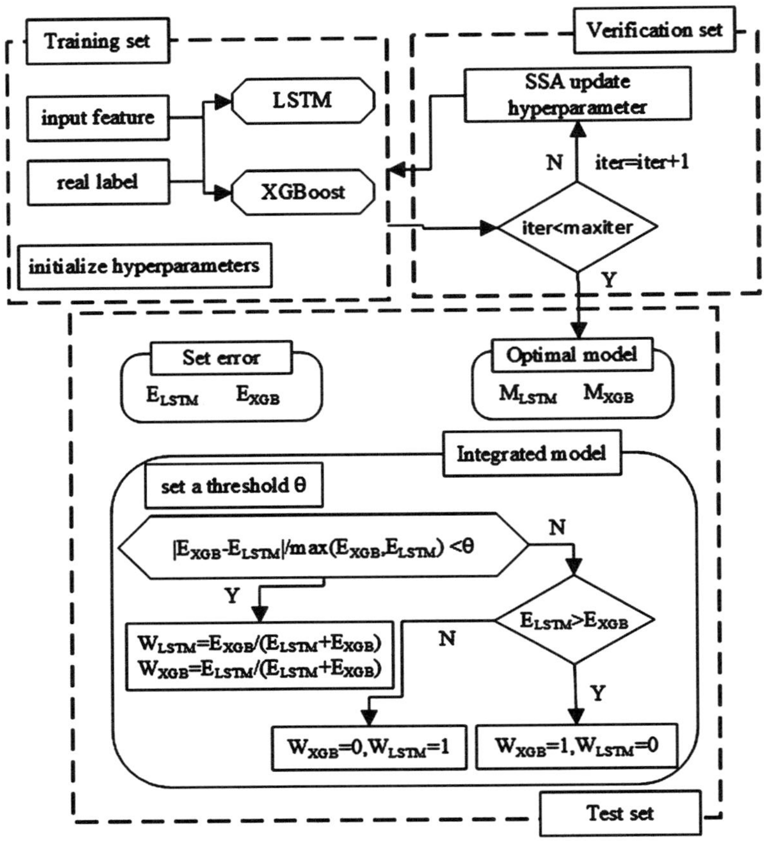

Fig. 1 LSTM-XGBoost model flowchart

Table 2 Feature importance

Number	Feature name	Feature importance
1	Land number	3.3220
2	Delay	1.5584
3	Real push time	1.0995
4	Plan takeoff time	1.0415
5	Queue length	0.8696

4.2 LSTM-XGBoost Results and Comparison

This section first determines the parameters of the optimized model and then determines the combination weight based on the prediction results of the optimized model.

After the LSTM model was optimized using the sparrow search algorithm, the model parameters were adjusted as follows: the number of neurons (LayerSizes) was 33, and the DropoutLayer was 0.1233.

The model parameters of XGBoost were adjusted as follows: the maximum number of iterations (Maxiter) was 99, the maximum depth (MaxDepth) was 3, and the minimum number of nodes (MinChild) was 3.

The combined weight of the finalized LSTM-XGBoost integrated model is: $W^{\text{LSTM}} = 0.51$, $W^{\text{XGB}} = 0.48$.

The predicted results, along with the corresponding scatter plots for gliding time predictions based on the optimized model parameters, are presented in Figs. 2 and 3. These visualizations provide a comprehensive overview of the model's predictive accuracy and performance under the refined parameter settings.

The prediction performance of the LSTM-XGBoost integrated model yielded a determination coefficient (R^2) of 0.81, a root-mean-square error (RMSE) of 3.34, and a mean absolute error (MAE) of 2.62. To further validate the superior predictive accuracy of the integrated model, the LSTM and XGBoost models were independently used to predict taxi-out times. Parameter optimization was performed using

Fig. 2 Predicted results diagram

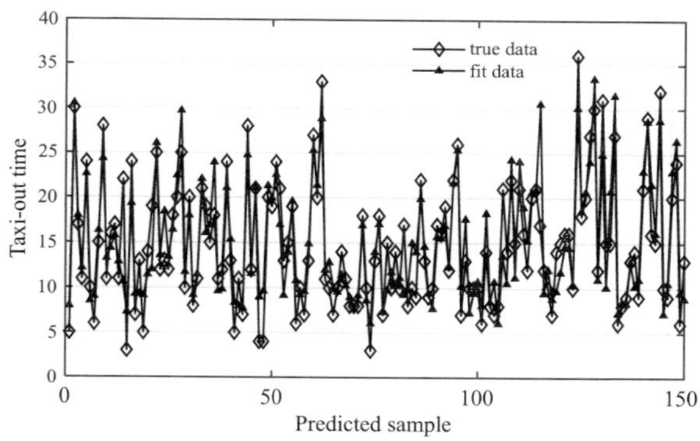

Fig. 3 Scatter diagram of results

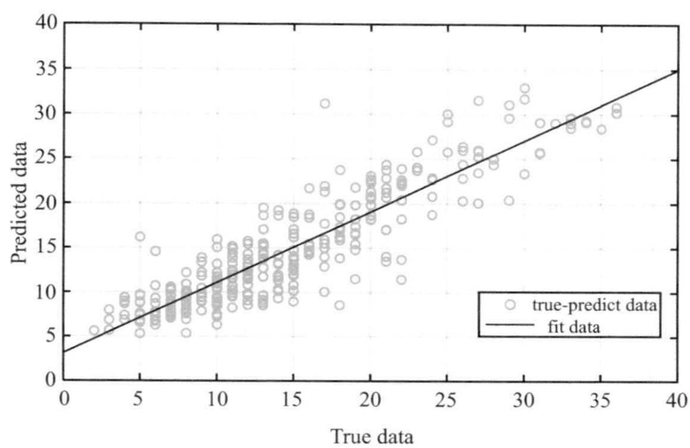

Table 3 Comparison of evaluation indicators

Index	Model		
	LSTM-XGBoost	LSTM	XGBoost
MAE	2.62	2.91	2.71
MSE	11.16	14.59	13.76
RMSE	3.34	3.82	3.71
R^2	0.81	0.73	0.77

the sparrow search algorithm, and the evaluation metrics for each standalone model were compared with those of the integrated model. The comparison results, presented in Table 3 and illustrated in Fig. 4, demonstrate the enhanced predictive performance of the combined approach.

Since the root-mean-square error (RMSE) is derived by taking the square root of the mean squared error (MSE), the trends of these two metrics are inherently aligned. However, RMSE provides a more precise measure of prediction accuracy compared to MSE. Consequently, the numerical trend of MSE is not presented separately in Fig. 4. From the figure, it is evident that the prediction errors—represented by the

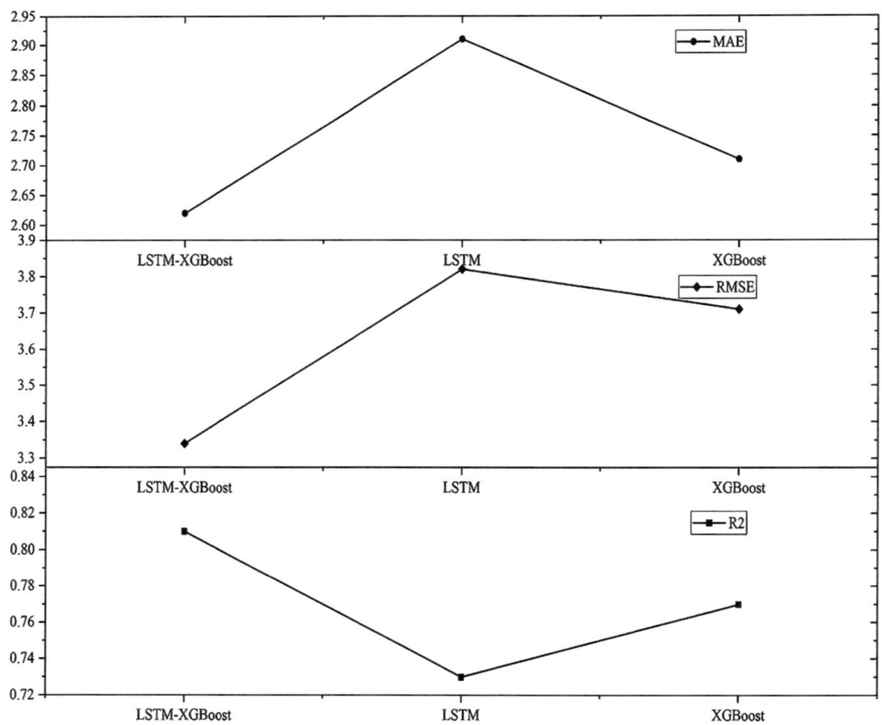

Fig. 4 Comparison of evaluation indicators

Table 4 Comparison of evaluation indicators (CGO)

Index	Model		
	LSTM-XGBoost	LSTM	XGBoost
MAE	3.12	3.91	3.57
MSE	13.99	18.58	17.14
RMSE	3.74	4.31	4.14
R^2	0.79	0.70	0.74

mean absolute error (MAE), MSE, and RMSE—are significantly smaller for the integrated LSTM-XGBoost model compared to the individual LSTM and XGBoost models. Specifically, the reduction in MSE error exceeds 3 min. Additionally, the coefficients of determination indicate that the integrated model's predictions are more accurate than those of the individual models.

To evaluate the generalizability of the proposed LSTM-XGBoost model, data from Zhengzhou Xinzheng International Airport (CGO) for April of the same year are selected for testing. Specifically, a total of 1842 departure records from April 15 to April 17 are used as the validation dataset. The evaluation metrics for the resulting predictions are presented in Table 4.

The Table 4 clearly shows the model error under the prediction using the data of Xinzheng International Airport. It can be seen that the overall model error is higher than the prediction error of the Capital Airport. Considering the different field conditions and scheduling capabilities of the two airports, the above error is acceptable. For each model, the prediction error of the LSTM-XGBoost model is still overall smaller than that of the single model, in which the MSE error time decreases by 3–4 min compared with the LSTM and XGBoost models, and the coefficient of determination is still much higher than that of the single model. Therefore, this study believes that the proposed integrated model can be better applied to various field conditions and has good generalizability.

The error analyses of the above models are based on the prediction results under normal weather conditions, while in the actual airport operation process weather factors have a greater impact on the operational efficiency of the field. When the prediction dataset contains weather data, especially the field data under abnormal weather conditions, the prediction accuracy of the model will have a certain impact.

To further examine the prediction accuracy of the proposed integrated model under varying time tolerance levels, the absolute error percentage accuracies for the LSTM-XGBoost and standalone LSTM and XGBoost models were calculated at $\pm$ 2, $\pm$ 3, and $\pm$ 5 min thresholds. These results are presented in Table 5, providing insight into the model's predictive reliability across different permissible error ranges.

In order to more intuitively observe the prediction accuracy of the three models at different tolerances at different times, this paper plots a comparison chart of the different models based on the percentage data in Table 5, as shown in Fig. 5.

The chart above shows the percentage of data points within $\pm$ 2, $\pm$ 3, and $\pm$ 5 min of the predicted value of each prediction model. It can visually reflect the prediction effect of each model within the error range. It is not difficult to see that the accuracy

Table 5 Absolute error percentage accuracy of each model

Tolerance	Model		
	LSTM-XGBoost (%)	LSTM (%)	XGBoost (%)
± 2 (min)	49.32	43.84	43.59
± 3 (min)	66.11	61.64	62.05
± 5 (min)	89.73	84.59	85.91

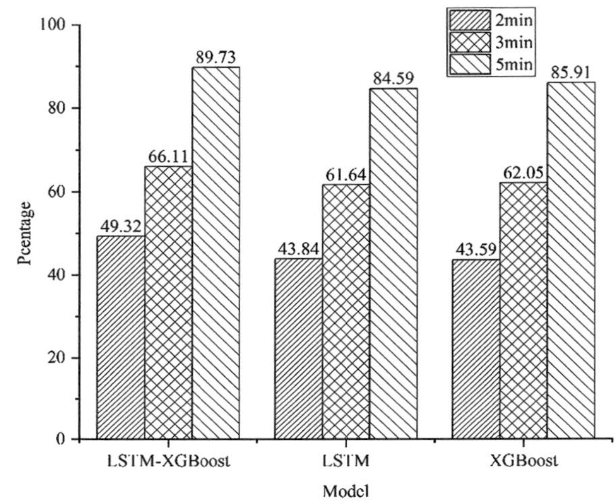

Fig. 5 Comparison of absolute percentage errors

of the LSTM-XGBoost model is higher than that of the two single models under the three time tolerances, and the accuracy under ± 5 min reaches 89.73%, which is much higher than that of the LSTM and XGBoost models. This study believes that the integrated model proposed by us can provide better guidance in actual scene operation management than a single model.

5 Conclusions and Future Research

Currently, most international departure taxiing time prediction approaches rely on single-model frameworks, leading to a plateau in achievable predictive accuracy. This study investigates the distinctive characteristics of the LSTM and XGBoost models and their respective strengths in regression tasks, proposing an integrated LSTM-XGBoost model to address this limitation. To enhance model performance, the sparrow search algorithm (SSA) is employed for hyperparameter optimization during training, leveraging the complementary advantages of both models to improve taxiing time prediction accuracy. Additionally, five feature parameters were identified from historical data at Capital Airport and analyzed for significance using a random forest model. Results indicate that each parameter exhibited an importance score exceeding 0.8, with four parameters scoring above 1.0, affirming that the selected model features are representative and highly relevant to the prediction task.

In conclusion, this study systematically compares the predictive performance of the integrated LSTM-XGBoost model with that of the standalone LSTM and XGBoost models using multiple evaluation metrics. The results consistently demonstrate that the integrated model outperforms the individual models across all assessed metrics. Moreover, the integrated model achieves superior prediction accuracy for varying time tolerances of $\pm$ 2, $\pm$ 3, and $\pm$ 5 min, confirming its robustness in predicting airplane departure taxiing times.

In conclusion, this study compares the error assessment metrics of the integrated LSTM-XGBoost model with those of the standalone LSTM and XGBoost models. The results demonstrate that the integrated model consistently outperforms the individual models in terms of prediction accuracy. Specifically, the integrated model shows higher accuracy than the standalone models at time tolerances of $\pm$ 2, $\pm$ 3, and $\pm$ 5 min, with its prediction accuracy approaching 90% at $\pm$ 5 min. This level of accuracy is deemed acceptable for real-world operations, where a 5-min tolerance is manageable due to variations in scheduling commands and aircraft taxiing speeds. These findings suggest that the LSTM-XGBoost integrated model offers superior performance in predicting aircraft departure taxi times. Furthermore, this model can be integrated into real-time airport traffic management systems (ATMS), utilizing live operational data (e.g., pushback time, current delays, queue length) as input to dynamically adjust taxi-out time predictions. Such a system would enable airport dispatchers to promptly identify delays, optimize runway allocation, and enhance ground handling scheduling, thereby reducing operational disruptions and improving overall field efficiency.

Given that the datasets used in this study consist solely of field data collected under normal weather conditions, it is not possible to effectively assess the impact of adverse weather phenomena, such as thunderstorms, blizzards, or fog, on the predictive performance of the model. Therefore, future research could incorporate meteorological variables—such as visibility, wind speed, precipitation, and snowfall—as input features for model training, along with their time-series data. Additionally, the integration of an attention mechanism could enable the model to dynamically focus on critical weather-related features. By analyzing model error metrics under both normal and abnormal weather conditions, the relative influence of these meteorological factors on model performance can be quantified. This analysis would facilitate further model optimization, enhancing its generalizability and robustness across varying environmental conditions.

Acknowledgements This work described in this paper is part of a research project funded by the China National College Students' Innovation Program Project Fund (No. 202210595058), the Guangxi Natural Science Youth Fund (No. 2021GXNSFBA075022), and the Natural Science Fund of China (No. 52462044). Their support is gratefully acknowledged.

References

1. Yin, J., Zhang, M., Ma, Y., Wu, W., Li, H., Chen, P.: Prediction and analysis of airport surface taxi time: classification, features, and methodology. Appl. Sci. **14**, 1306 (2024)
2. Qiao, P., Hu, M., Yin, J., Su, J., Chen, Y., Yin, M.: An adaptive similar scenario matching method for predicting aircraft taxiing time. Aerospace **11**, 461 (2024)
3. Lim, Y., Alam, S., Tan, F., Lilith, N.: Variable taxi-out time prediction based on machine learning with interpretable attributes. In: *Transactions Of The Japan Societyfor Aeronautical And Space Sciences*, vol. 67(Issue 3), pp. 136–144 (2024) Released on J-STAGE May 04, 2024, Online ISSN 2189-4205, Print ISSN 0549-3811
4. Herrema, F., Curran, R., Visser, H., Huet, D., Lacote, R.: Taxi-out time prediction model at Charles de Gaulle airport. J. Aerosp. Inf. Syst. **15**, 3, 120–130 (2018)
5. Lian, G., Zhang, Y., Desai, J., Xing, Z., Luo, X.: Predicting taxi-out time at congested airports with optimization-based support vector regression methods. Math. Probl. Eng. **2018**, Article ID 7509508, 11 (2018).
6. Chen, J., Ravizza, S., Atkin, J.A.D. et al.: On the Utilisation of Fuzzy Rule-Based Systems for Taxi-out time Estimations at Airports (2011)
7. Kim, J., Baik, H.: Taxi-out time prediction at a busy airport using random forest algorithm. In: 2021 IEEE/AIAA 40th Digital Avionics Systems Conference (DASC), San Antonio, TX, USA (2021)
8. Okwir, S., Amouzgar, K., Ng, A.H.C.: Exploring prediction accuracy for optimal taxi times in airport operations using various machine learning models. J. Air Transp. Manag. **122**, 102684 (2025). ISSN 0969-6997.
9. Tang, X., Ye, M., Zhang, S., Fuellhart, K.: Prediction of taxi-in time and analysis of influencing factors for arrival flights at airport with a decentralised terminal layout. Promet—Traffic Transp. **36**(4), 623–638 (2024)
10. Yang, X., Yang, H., Mao, Y., Wang, Q., Yin, S.: Multi-task dynamic spatio-temporal graph attention network: a variable taxi-out time prediction model for airport surface operation. Aerospace **11**, 371 (2024)
11. Park, D.K., Kim, J.K.: Influential factors to aircraft taxi-out time in airport. J. Air Transp. Manag. **106**, 102321 (2023). ISSN 0969-6997

Research on China Freight Volume Forecasting Based on LSTM

Quanli Mo, Liwei Li, Kugan Huang, Yanfang Pan, Shiyu Wang, Bote Liu, and Guobin Gu

Abstract With the rapid development of China's economy, freight volume, as a crucial indicator in transportation and supply chain management, holds significant importance in optimizing resource allocation and policymaking. To accurately forecast trends in freight volume changes, this paper proposes a deep learning model based on long short-term memory (LSTM) networks, incorporating Bayesian optimization to adjust the model's hyperparameters, thereby enhancing forecasting performance. During the data preprocessing phase, the linear interpolation method is used to address data missing issues, ensuring the integrity and continuity of the time series data. The results demonstrate that the LSTM model based on Bayesian optimization exhibits high accuracy and stability in predicting freight volume time series. The paper provides a methodology for forecasting complex time series data and offers direction for research on data prediction problems in transportation and supply chain management concerning freight volume forecasting accuracy.

Keywords LSTM · Freight volume forecasting · Bayesian optimization · Deep learning

1 Introduction

China's national freight volume data is crucial to economic activities, as it not only reflects the state of the national economy but is also used to assess and forecast economic growth. Accurate freight volume forecasts provide fundamental data support for freight companies and logistics enterprises in transportation planning and resource allocation. Therefore, it is essential to conduct in-depth research on forecasting methods for China's freight volume data. A reasonable and effective freight volume data processing method can help enterprises make more agile decisions in a dynamic market environment, optimize resource allocation, and improve transportation efficiency, thereby reducing operational costs and enhancing competitiveness

Q. Mo · L. Li · K. Huang · Y. Pan · S. Wang · B. Liu · G. Gu (✉)
College of Traffic and Transportation, Nanning University, Nanning, Guangxi, China
e-mail: 15251707205@163.com

K. Subramaniyam and M. Paprzycki (eds.), *Smart Cities and the Internet of Things*, Internet of Things, https://doi.org/10.1007/978-3-032-08290-9_4

in the industry. In the study of freight volume forecasting, methods can be broadly divided into two categories: statistical model methods and machine learning methods. Ni et al. [1] used a grey prediction model to handle uncertainty and incompleteness in data, providing relatively reliable forecasts for multimodal freight volume in Henan Province. Cao et al. [2] proposed a railway freight volume forecasting method based on a decomposition-ensemble framework, predicting with a new dataset composed of recombined components and principal components through an extreme learning machine model optimized by the bald eagle search algorithm. Tan et al. [3] used deep neural networks to establish single-step and multistep forecasting models for short-term railway freight demand prediction. Akter et al. [4] utilized large-scale Global Positioning System (GPS) data for freight planning, providing insights into trip chaining behavior and filling critical data gaps that currently limit the development of advanced freight forecasting models. Tatarintsev et al. [5] determined the impact of various macroeconomic indicators from different regions worldwide on freight volume between specified areas, identifying patterns between regions and applying necessary mathematical models for prediction. Snoek et al. [6] proposed a Bayesian optimization method that effectively enhances the hyperparameter tuning efficiency of machine learning models, offering new ideas for forecasting complex time series data. These studies provide theoretical support and an international perspective for the design and optimization of the freight volume forecasting model in this paper, which uses a long short-term memory (LSTM) model based on Bayesian optimization to study the prediction of China's freight volume time series data.

2 Methodology

2.1 ALSTM Principles

Long short-term memory (LSTM) networks are a special type of recurrent neural network (RNN) specifically designed to handle and predict time-series data. Compared to traditional RNNs, LSTM networks contain memory cells that can better capture long-range dependencies, preventing the loss of information in long sequences. The structure of a memory cell is shown in Fig. 1.

Long short-term memory (LSTM) models have long been regarded as an excellent algorithm for time series data forecasting, being a variant of the recurrent neural network (RNN) algorithm. The gated mechanism introduced by LSTM can effectively alleviate the vanishing gradient problem, allowing it to handle longer sequence data [7]. LSTM is particularly suitable for dealing with events where there are very long intervals and delays between events, such as the research on filling missing data in China's freight volume discussed in this paper.

The LSTM module's units primarily consist of a forget gate, an input gate, a cell state, and an output gate, with the following calculation formulas:

The forget gate calculation formula is:

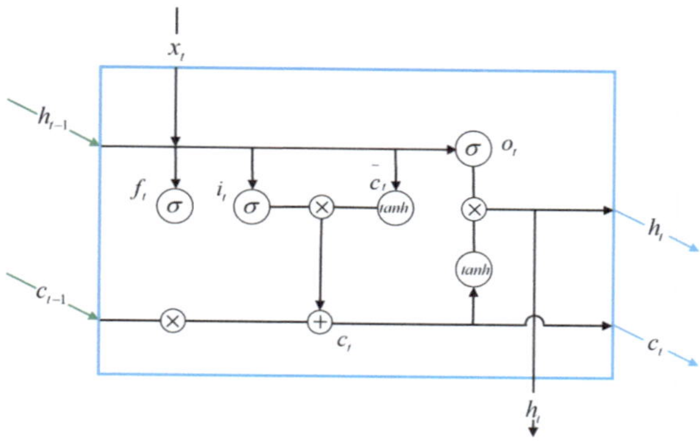

Fig. 1 Long short-term memory structure diagram

$$f_t = \sigma(W_f x_t + U_f h_{t-1} + V_f c_{t-1} + b_f) \tag{1}$$

The input gate calculation formula is:

$$i = \sigma(W_i x_t + U_i h_{t-1} + V_i c_{t-1} + b_i) \tag{2}$$

The output gate calculation formula is:

$$\bar{c_t} = \tanh(W_c x_t + U_c h_{t-1} + b_c) \tag{3}$$

$$c_t = f_t \odot c_{t-1} + i_t \odot \bar{c_t} \tag{4}$$

$$o_t = \sigma(W_o x_t + U_o h_{t-1} + V_o c_t + b_o) \tag{5}$$

$$h_t = o_t \odot \tanh(c_t) \tag{6}$$

Among them, given the original input x_t, the input gate is i_t, the output gate is o_t, the forget gate is f_t, the cell state is c_t, the final output is h_t, t is the time step, $\odot$ is the Hadamard product, and the initial values are set to $c_t = 0$, $h_t = 0$. W, U, and V are weight matrices, and b represents the bias vector parameters.

$$\sigma(x) = \frac{1}{1 + e^{-x}} \tag{7}$$

$$\tanh(x) = \frac{e^x - e^{-x}}{e^x + e^{-x}} \tag{8}$$

Among them, $\sigma(x)$ represents the sigmoid activation function, and $\tanh(x)$ represents the hyperbolic tangent activation function.

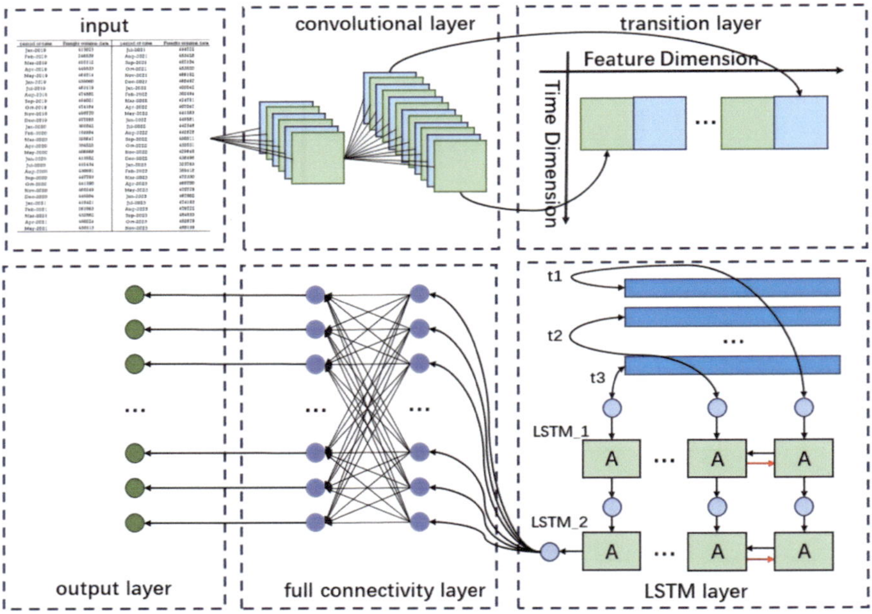

Fig. 2 Bi-LSTM architecture diagram

2.2 Bi-LSTM Principles

Bidirectional-LSTM is an extended LSTM model that introduces a backward LSTM layer into the network structure, allowing the model to capture both past and future sequence information simultaneously and more comprehensively capture the contextual information in sequence data [8]. By combining two independent LSTM networks to process information in both the forward and backward directions of the sequence, the model can consider contextual information from both sides, enhancing its ability to understand time series or text data. Its structure is shown in Fig. 2.

3 Experiments

3.1 Data Sources

The nationwide freight volume data in China is sourced from the National Bureau of Statistics of China. Due to missing data in the national database, the integrity of the time series is affected. Therefore, when extracting complete and continuous time-series data, effective missing-value handling methods need to be used to ensure the accuracy and reliability of the data. The paper extracted nationwide freight volume data from January 2015 to December 2022 from the national data as the dataset, with a sample interval of 1 month, totaling 96 data entries. Examples of some data are shown in Table 1.

By cleaning and preprocessing the freight volume data of China, anomalies and errors in the dataset were identified and corrected to ensure the quality and reliability of the data. For handling missing values, the paper used the linear interpolation method to fill in the missing values of the nationwide freight volume. Linear interpolation is a simple and effective filling strategy suitable for time series data, allowing the calculation of missing values based on data from adjacent time points, thus maintaining data continuity and trend.

Due to the significant impact of the COVID-19 pandemic in 2020 on China's freight volume, which resulted in abnormal fluctuations in the data for that year, it was decided to exclude the 2020 nationwide freight volume data to avoid interference

Table 1 Nationwide freight volume data (portion)

Period of time	Freight volume data	Period of time	Freight volume data
Jan 2018	386,354	Jan 2021	415,421
Feb 2018	244,273	Feb 2021	261,863
Mar 2018	391,601	Mar 2021	432,862
Apr 2018	423,516	Apr 2021	450,224
May 2018	440,203	May 2021	458,113
Jun 2018	434,978	Jun 2021	456,678
Jul 2018	433,435	Jul 2021	444,772
Aug 2018	450,743	Aug 2021	453,428
Sep 2018	468,951	Sep 2021	457,104
Oct 2018	462,415	Oct 2021	453,520
Nov 2018	472,266	Nov 2021	466,152
Dec 2018	442,669.5	Dec 2021	460,497
Jan 2019	413,073	Jan 2022	402,042
Feb 2019	246,539	Feb 2022	302,494
Mar 2019	425,112	Mar 2022	424,751
Apr 2019	449,533	Apr 2022	407,047
May 2019	464,814	May 2022	441,383
Jun 2019	459,060	Jun 2022	449,581
Jul 2019	463,119	Jul 2022	442,348
Aug 2019	474,882	Aug 2022	442,878
Sep 2019	494,021	Sep 2022	450,511
Oct 2019	474,184	Oct 2022	432,031
Nov 2019	498,770	Nov 2022	429,645
Dec 2019	477,265	Dec 2022	436,496

from the extreme conditions caused by the pandemic on the overall analysis results. This processing ensures that the dataset used is more stable and reflects the freight volume trend under normal circumstances.

The paper used 48 data entries from the processed nationwide freight volume dataset as the training set and the remaining 12 data entries as the test set. In the constructed LSTM model, by iteratively predicting the values for the next 12 time points, the final prediction sequence corresponding to the test set was obtained.

3.2 Set Parameters and Train the Network

When building deep neural networks, the selection of hyperparameters has a crucial impact on the model's performance. In the paper, Matrix Laboratory (MATLAB) 2023b was used to build the deep neural network, and Bayesian optimization was chosen for hyperparameter optimization of the two LSTM models. The optimization mainly focused on the number of network layers, the number of units per layer, and the learning rate. The specific range of values for each hyperparameter is shown in Table 2. To prevent the model from overfitting and improve its generalization ability, a dropout layer is added to the constructed LSTM network with a probability of 50%. Furthermore, the sigmoid function is selected as the activation function of the forgetting gate, the input gate, and the output gate, while the Tanh function is used in generating the candidate memories as the activation function.

To find the optimal hyperparameter configuration and improve the model's performance, the experiment used Bayesian optimization to pretrain the network with training samples. Bayesian optimization probabilistically models the hyperparameters and intelligently selects the next evaluation point to maximize the model's performance.

During the pretraining phase, the Bayesian optimization algorithm continuously explores the hyperparameter space by selecting different combinations of hyperparameters to train the model, constantly evaluating its performance on the training set. After pretraining the network with the training samples, the range of hyperparameters for the two networks is shown in Table 3. The relationship between the number

Table 2 Hyperparameter value range

Hyperparameter	Range of values
Time lag	12
Solvers	Adam optimizer
Training epoch	[1400]
Number of layers	[1,2]
Number of units	[50,200]
Use Bi-LSTM layer	[0,2]
Initial learn rate	[0.01,1]

Table 3 Network hyperparameter range

Number of layers	Number of units	Use Bi-LSTM layer	Initial learn rate	Regularization factor
1	141	1	0.6109	0.0071

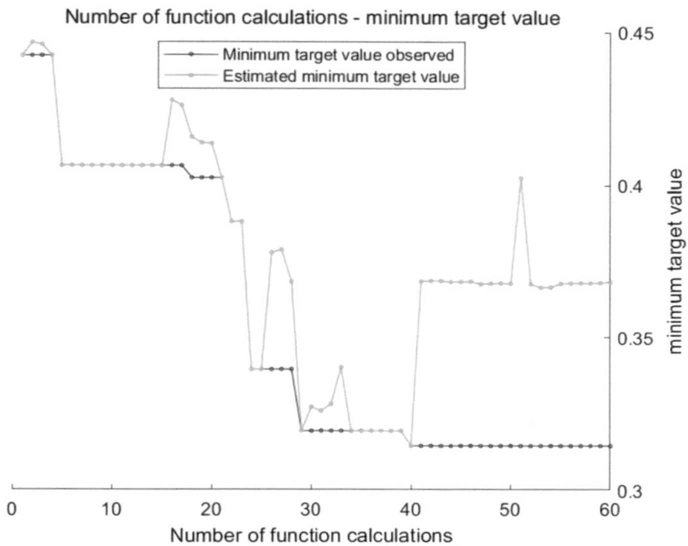

Fig. 3 Number of function calculations–minimum target value

of function evaluations during training and the minimum objective value is depicted in Fig. 3.

4 Results

Based on the parameter results of Bayesian optimization, the results of each part of the model constructed for China's freight volume prediction are shown in Fig. 4. The correlation coefficient (R^2) and root mean square error (RMSE) are used as evaluation metrics for the prediction accuracy of China's freight volume.

As seen in Table 4, the accuracy of the model's total sample, training set, and prediction set is basically the same, indicating that the model has good generalization ability. However, the model has a weaker ability to capture outliers (i.e., months with abrupt changes in freight volume). The COVID-19 pandemic significantly affected China's freight volume, leading to a marked increase in the volatility and uncertainty of the time series data.

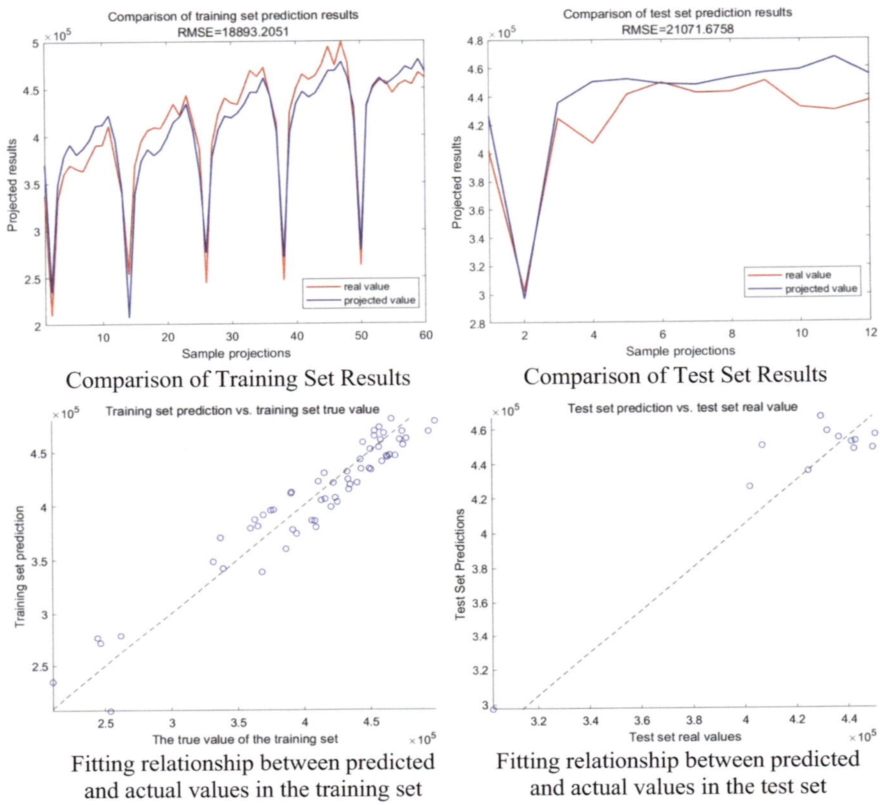

Fig. 4 Fitted regression plot

Table 4 Error analysis

Model	R^2	RMSE
Training set	0.91276	16,809.9434
Test set	0.70528	16,585.9688

After excluding the data from the pandemic period, the remaining time series data became sparser, leading to an insufficient data sample, which in turn affected the model's training and predictive capabilities. The reduction in sample size may result in the model being unable to fully learn the underlying patterns and rules in the data, thereby reducing the accuracy and reliability of the predictions. Moreover, the scarcity of data also means that the model lacks sufficient capability to cope with potential future anomalies.

5 Conclusion

The paper proposes an improved freight volume forecasting model by combining the LSTM network with the Bayesian optimization algorithm. This model effectively addresses the issues of trend capture and missing data in time series forecasting, demonstrating good fitting ability and forecasting accuracy. However, the model's adaptability in data scarcity and extreme events remains limited, particularly in the context of unexpected events such as the new crown epidemic, where its prediction performance is constrained.

In the future, with the continuous accumulation of freight volume observation data and the continuous optimization of related algorithms, the model performance can be further improved in several ways. These include combining multivariate analysis with information from external variables, developing a more robust deep learning framework, and enhancing real-time prediction. The thesis proposes several avenues for further improvement, including the utilization of data augmentation techniques to address data scarcity, cross-domain algorithmic migrations, and enhanced interpretability. These enhancements have the potential to enhance the accuracy of the thesis's predictions, thereby providing theoretical support and practical reference for the optimization of transportation and supply chain management.

Acknowledgements A Study on the Construction of Case Bank for Civics Teaching in Applied Undergraduate Logistics Engineering Practice Class Courses (2024XJJG23).

References

1. Niu, L., Ge, J.: Research on the prediction of multimodal transport freight volume in henan province based on grey forecasting model. J. Logistics Technology **47**(17), 88–91 (2024)
2. Cao, H., Qin, J.: Research on railway freight volume prediction method based on decomposition-integration framework. J. Comput. Technol. Dev. **33**(08), 192–198 (2023)
3. Tan, X., Zhang, X.: Short-term railway freight volume prediction based on GRU deep network. J. J. China Railw. Soc. **42**(12), 28–35 (2020)
4. Akter, T., Hernandez, S., Camargo, P.V.: Freight operational characteristics mined from anonymous mobile sensor data. J. Transp. Res. Rec. **2677**(9), 236–247 (2023)
5. Tatarintsev, M., Nikitin, P., Korchagin, S.: Computer modeling of the dynamics of interregional freight transport depending on macroeconomic indicators. C. ICAART **2**, 1249–1257 (2021)
6. Snoek, J., Larochelle, H., Adams, R.P.: Practical bayesian optimization of machine learning algorithms. J. Adv. Neural Inf. Process. Syst. **25** (2012)
7. Cao, L., Xie, W., Qin, D., et al.: Research on missing data imputation for bridge construction monitoring based on LSTM. J. Highw. (11), 217–223 (2024)
8. Jin, N., Jiang, H.: Logistics review sentiment analysis based on BiLSTM-CNN-MultiHead attention-dropout. J. Logist. Technol. **46**(23), 48–52 (2023)

Urban Flood Simulation and Resilience Assessment Based on Cellular Automata and System Performance

Yuelei Wu, Rupeng Zhao, Qingguo Li, and Yang Chen

Abstract The CADDIES model, based on Cellular Automata, is adopted to simulate the flood routing and inundation under the observed storm rainfall events and the design rainfall process in the Jinan urban area. Then, with a resilience assessment method based on a system performance method, the flood resilience in each sub-basin of the study area is calculated. The results show that the direction of flooding expansion is mainly from the south to the north, and inundation occurs more frequently in the northern and central regions. The main impact factors of urban resilience are land use, elevation, and average land slope. The spatial distribution of the flood resilience of the Jinan urban area is similar to the flood growth trends. This study clarifies the impact mechanism of urban rainstorms and flood disasters on urban recovery, which can improve urban disaster prevention and mitigation capacity and promote regional sustainable development.

Keywords Jinan urban area · Cellular automata · CADDIES model · Flood simulation

1 Introduction

In recent years, rapid urbanization and climate change have intensified the urban surface flood [1], which has caused frequent occurrences of urban waterlogging disasters and posed a huge threat to the safety of urban infrastructure, human lives, and property in many cities all over the world [2]. As the capital city of Shandong Province, Jinan assumes the pivotal role as the economic and cultural hub of Shandong Province, thereby playing a crucial part in driving its economic and social development. However, Jinan is plagued by recurrent floods, exemplified by three

Y. Wu · Q. Li (✉) · Y. Chen
School of Water Conservancy and Environment, University of Jinan, Jinan, China
e-mail: stu_liqg@126.com

R. Zhao
Shandong Provincial Hydrological Center, Jinan, China

K. Subramaniyam and M. Paprzycki (eds.), *Smart Cities and the Internet of Things*, Internet of Things, https://doi.org/10.1007/978-3-032-08290-9_5

significant rainstorms and inundations on July 16, 1962; August 26, 1987; and July 18, 2007. This flood calamity has emerged as one of the most devastating natural disasters in Jinan's history. Urban flood simulations are an important nonengineering measure of analyzing urban flood characteristics and evaluating them [3]. Based on the cellular automata (CA) Dual-DraInagE Simulation (CADDIES) model framework combined with the improved CAFloodPRO program, an urban flood simulation model was constructed using data from different resolutions, such as urban boundaries, land use conditions, road information, and water system characteristics, to simulate the flood situation in the Jinan urban area under various conditions. We then evaluated the recovery capacity of Jinan's flood disasters [4]. The focus lies in addressing the challenges of urban rainstorm and flood simulation, as well as evaluating urban flood resilience. This entails quantitatively unveiling the spatio-temporal response patterns between urban floods and flood resilience, resolving concerns regarding focal areas and regions within urban flood ecological management plans, while also facilitating optimal utilization of urban flood drainage capacity and rainstorm/flood disaster ecological management. Consequently, this enhances the city's disaster prevention and reduction capabilities, promotes new urbanization construction, and fosters regional sustainable development.

2 Overview of the Study Area

Jinan is located in the central-western region of Shandong Province, between 36°01′ and 37°32′ N and 116°11′ and 117°44′ E, with an area of approximately 10,244 km^2, as shown in Fig. 1. The main urban area is the main research object of this study and involves six districts, including the Lixia and Shizhong districts, with an area of approximately 424 km^2. To explore the spatial variation of flood inundation and resilience, the study area is divided into 28 subcatchments, which are identified from zone C1 to C22, also shown in Fig. 1.

3 Materials and Methods

The CADDIES model was developed based on the principle of cellular automata (CA) [5, 6], and achieves computational accuracy through simple water volume change rules. Existing computable models in the CafloodPro program include the Weighted Cellular Automata 2D model version 1 (WCA2Dv1) and Weighted Cellular Automata 2D model version 2 (WCA2Dv2). In the earlier version of WCA2Dv1, there was a phenomenon of water level oscillation in adjacent cells during the calculation process, which caused instability and significant errors in the simulation process; as a result, the model could not meet the requirements of parallel computation on various grids, resulting in longer computation time in time steps, which seriously affects the efficiency of model operations. The subsequent versions of WCA2Dv2

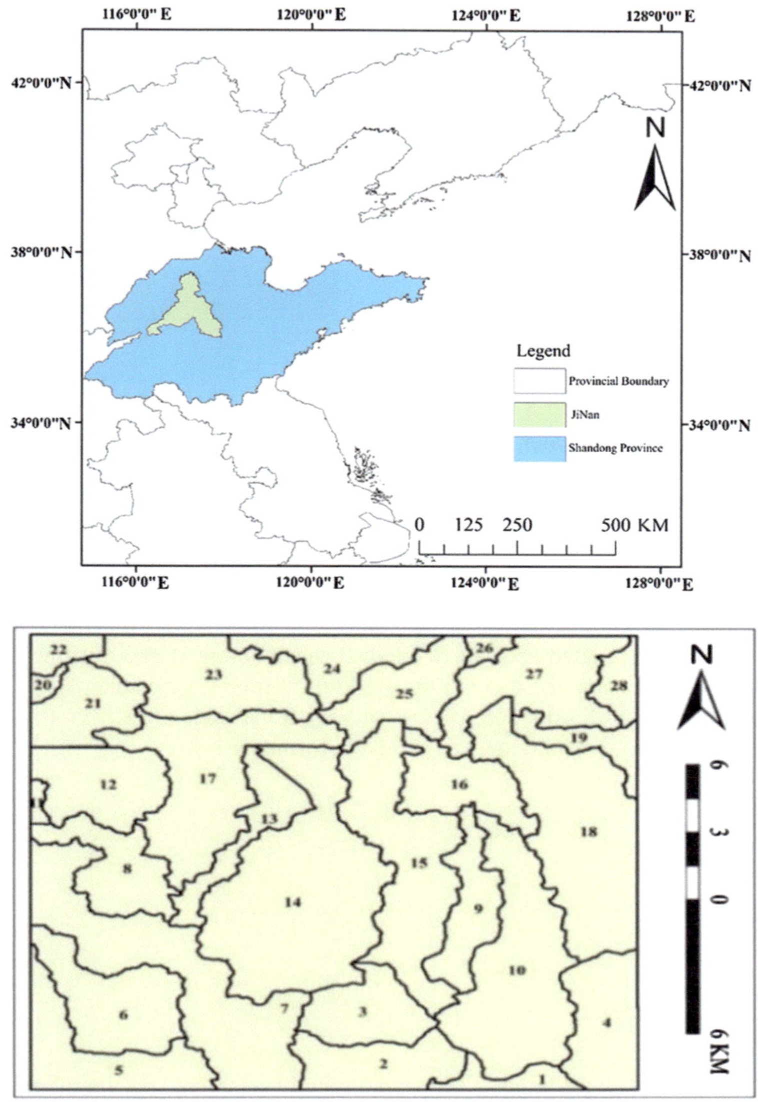

Fig. 1 Location of the study area and catchment distribution

were mainly based on the square discretization of John von Neumann-related neighborhoods. The main steps include: ① identify grid neighborhood cells; ② calculate the weight ratio of each downstream grid based on the water flow accumulation situation; ③ analyze and calculate the total water output of the central grid; and ④ noting that the water output between downstream grids consists of two parts: weight

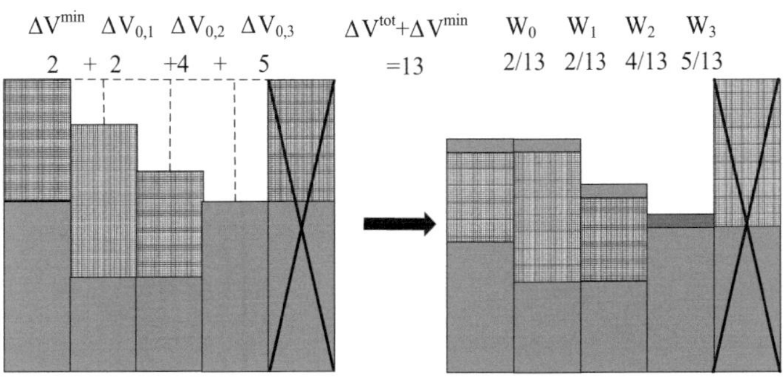

Fig. 2 Weight calculation and water exchange chart

and total transmission water, obtain the final value by parallel calculation of each element [7] (Fig. 2).

The system performance function curve of the area within the system is mainly evaluated by the presence or absence of an inundation area. When water depth inundation occurs, it indicates damage to urban functions, causing economic losses and interfering with human activities. The ratio of submerged area to total area is used to reflect the strength of resilience. When attacked by a rainstorm, the grid state value shows two states: flooded and not flooded. If the submerged depth of the ith grid is greater than the submerged depth threshold h_c at time t, the value of $g(i, t)$ is 1, indicating that the grid is flooded. On the contrary, if the value of $g(i, t)$ is 0, it means that the grid has not been flooded, as shown in formulas 1–4.

$$g(i,t) = \begin{cases} 0 d(i,t) \le h_c \\ 1 d(i,t) \ge h_c \end{cases} \tag{1}$$

$$N(t) = \sum_{i=1}^{N} g(i,t) \tag{2}$$

$$p(t) = 1 - \frac{N(t)}{N} \tag{3}$$

where $g(i, t)$ is the ith grid state at time t in the study area; $d(i, t)$ is the submergence depth at time t in the region; h_c is the threshold for submerged water depth; $N(t)$ is the total number of grids in the region; $p(t)$ is the system performance;

$$\text{Res} = \frac{1}{t_2 - t_1} \int_{t_1}^{t_2} p(t)dt \tag{4}$$

where Res is the flood resilience; t_1 is the initial time of occurrence of water depth submergence; t_2 is the five-hour time when the initial rainstorm starts. Considering the setting of the calculation periods t_1 and t_2, based on the relevant research results of urban flood simulation [8], the most representative and reflecting grid response mechanism calculation periods were considered. It is believed that the flood resilience is influenced by factors such as pregnancy factors, and the time period during which the flood is most severely affected is considered the most objective factor to reflect the actual results of flood recovery ability evaluation.

4 Results and Discussion

4.1 Multi-Scenario Flood Simulation Based on CADDIES

Rainfall for the storm of August 26, 1987, in Jinan, numbered “870,826,” is used as the input rainfall pattern to simulate the changes in the submerged depth of the urban area of Jinan at different time points during this storm, as shown in Fig. 3.

As shown in Fig. 3, from the beginning to the end of rainfall, the phenomenon of “flooding from south to north” is obvious. The central and northern regions have a longer confluence time, a large inundation area, and flood subsides slowly, resulting in a slower recovery capacity. The phenomenon of road flooding was relatively

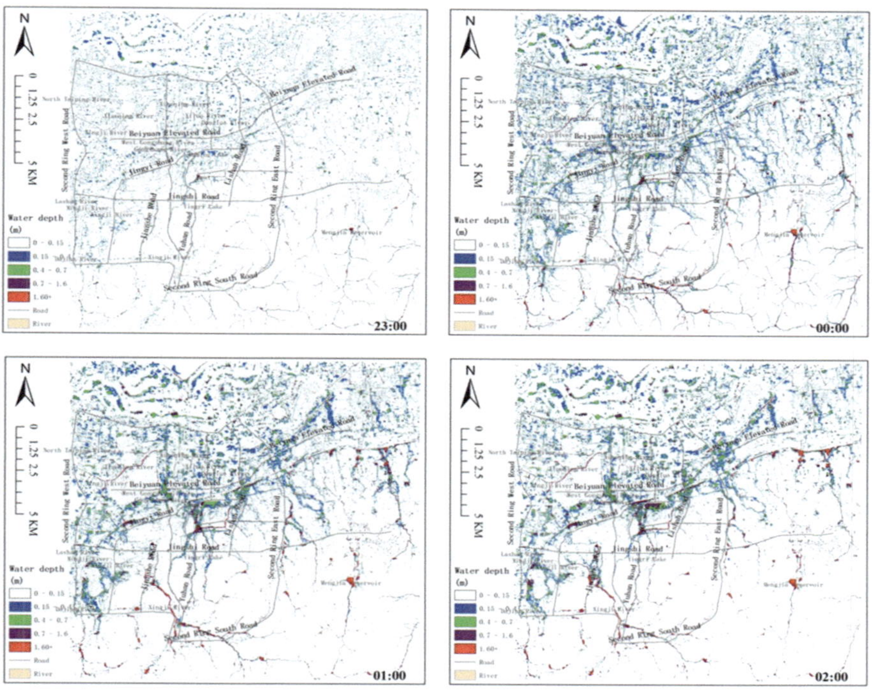

Fig. 3 Simulating inundated depths with input rainfall for the storm “870,826”

pronounced and concentrated in areas such as Second Ring East Road, Yuhan Road, and Yingxiongshan Road. The drainage of the Xiaoqing River, Xingji River, and surrounding moats takes a long time; water depth and inundated area increase, the pressure of the drainage system is high, and waterlogging problems are prominent.

Considering the increasing intensity and frequency of urban extreme rainfall events due to global warming, the maximum return period was set to 200 years. The rainfall intensity hydrograph under the return period of 200 years is obtained from the combination of the Chicago rain-type formula and the Jinan rainstorm intensity formula, as shown in Fig. 4.

4.2 Maintaining the Integrity of the Specifications

The resilience of flood disasters in each catchment area is calculated using formula 4, as shown in Fig. 5. Catchment areas such as C1 and C2, located in higher-elevation areas in the southern mountainous areas, have obvious regional advantages in terms of flood resilience. The 13 catchment areas distributed outside the main urban area, such as C5, C6, and C7, have relatively strong resilience in flood disasters due to their strong infiltration and drainage capacity, relying on advantages such as river channel flood discharge and their own slope; however, the 10 catchment areas located in the main urban area, including C12 and C13, as well as the catchment area C8 located in the lower reaches of the Xingji River, are affected by the convergence of rainwater and aquatic products in adjacent catchment areas, land use factors, rainfall types, and distribution characteristics of pipeline networks, resulting in weak infiltration capacity and overall weak resilience of flood disasters. The catchment C27 is affected by the Xiaoqing River and other watersheds, and the maximum water depth exceeding 1.2 m is located near the water body. Because of the strong flood discharge capacity near the watershed, its flood disaster resilience is relatively strong.

Through a resilience assessment of each catchment area with a 200-year return period, the main disaster-causing factor for the occurrence of inundation grid changes in each catchment area was found to be the inundation depth threshold. The transport characteristics of the overall submerged area were analyzed based on the changes in water depth at different times within 5 h after a 200-year recurrence interval of 2 h, and C10, C14, C17, and C27 were selected for typical analyses, as shown in Fig. 4. The submerged area of the catchment area C14 is relatively large, and the inundation range widens with the increase of rainfall time; the inundation range of C10 exceeding 0.15 m begins to decrease after 2 h; the elasticity indices of catchment areas C10 and C18 are 0.892 and 0.731, indicating that when suffering from urban flood disasters, catchment area C10 has a shorter flood duration and stronger water withdrawal ability, and can quickly recover to its initial state. The maximum water depth and inundation range of catchment area C17 do not change significantly within 5 h, but the speed of flood retreat is relatively slow.

Fig. 4 The maximum submergence depth for a 200-year recurrence interval of 1 h, 2 h, and 3 h

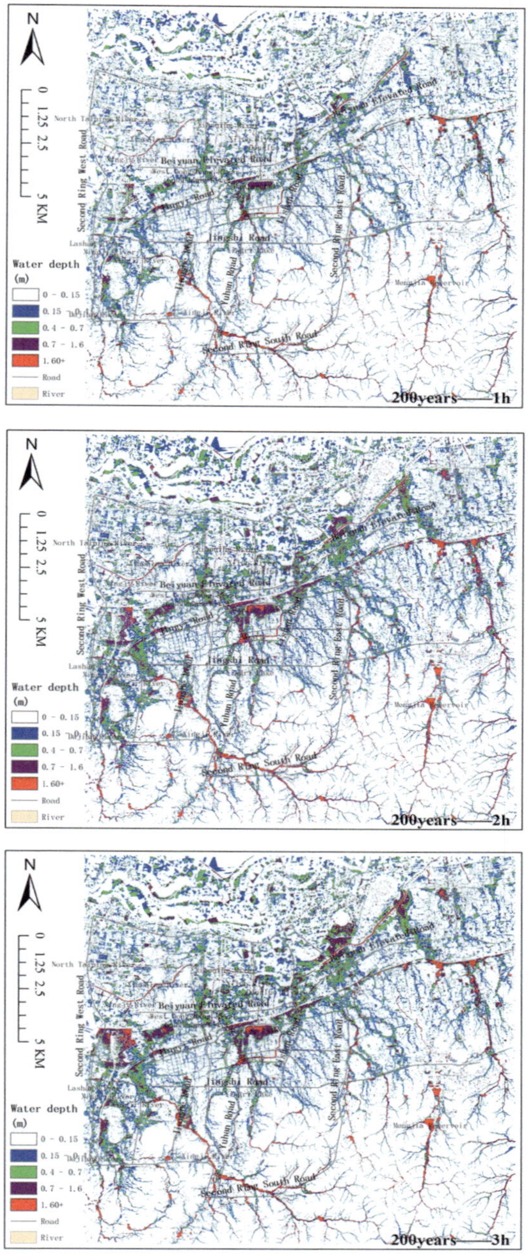

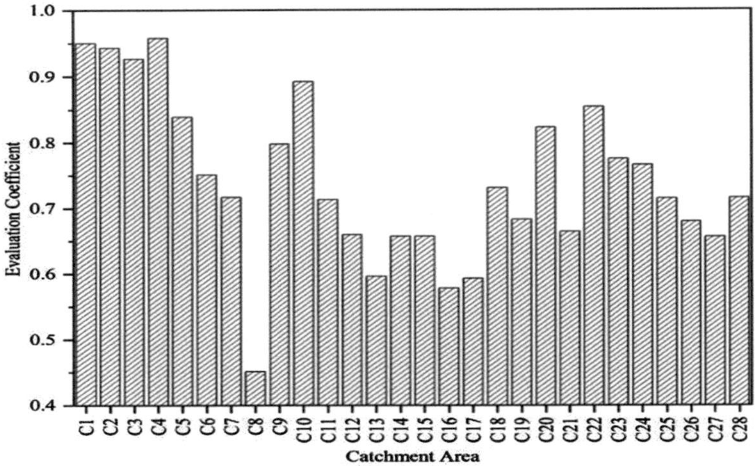

Fig. 5 Flood disaster resilience in different catchments

5 Conclusions

(1) Through the combination of actual rainfall patterns and design rainfall patterns, the CADDIES model can better simulate the process of flood inundation and migration in Jinan city.

(2) The catchment area with higher elevation in the southern mountainous area has obvious regional advantages in flood disaster resilience; the catchment area distributed outside the central urban area relies on the advantages of wider land use area such as grassland and its average slope, and the flood disaster resilience is high; the resilience of the catchment area located in the central urban area and the urban construction land area is overall lower.

Acknowledgements This work was supported by the Shandong Province Natural Science Foundation Project (Ref: ZR2021ME229).

References

1. Zhang, J.Y., Song, X.M., Wang, G.Q., et al.: Development and challenges of urban hydrology in a changing environment. Adv. Water Sci. 594–605 (2014)
2. Hammond, M.J., Chen, A.S., Djordjević, S., Butler, D., Mark, O.: Urban flood impact assessment: a state-of-the-art review. Urban Water J. **12**(1), 14–29 (2015)
3. Li, N.: Application of urban flood simulation technology in urban flood management. China Flood Drought Manag. 5–6 (2019)
4. Liao, K.H., Lin, H.J., Wang, Y.: A theory on urban resilience to floods: a basis for alternative planning practices. Urban Plan. Int. 388–395 (2015)

5. Wang, Y., Liu, Z., Wang, G., Xue, W.: Cellular automata based framework for evaluating mitigation strategies of sponge city. Sci. Total. Environ. **796**, 148991 (2021)
6. He, M., Chen, C., Zheng, F., Chen, Q., Zhang, J., Yan, H., Lin, Y.: An efficient dynamic route optimization for urban flooding evacuation based on cellular automata. Comput. Environ. Urban Syst. **87**, 101622 (2021)
7. Chang, X.D., Xu, Z.X., Zhao, G., et al.: Spatiotemporal variations in precipitation in Ji'nan. J. Beijing Norm. Univ. (Nat. Sci.) **53**(05), 567–574 (2017)
8. Sherman, L.R.K.: Founder of 'unit line Method. Hebei Water Resour. 28 (2017)

Research on the Scale Calculation Model of Port Logistics Park

Shiwen Sun, Peng Cai, and Jinxiu Fu

Abstract Logistics parks are the inevitable outcome of modern logistics industry development to a certain stage. Rational planning and effective use of logistics park land can improve the operational efficiency of ports, reduce logistics costs, and enhance the competitiveness of ports. Port logistics parks play a significant role in improving port functions, enhancing the level of port operation organization, and expanding the port's radiation service capacity. Due to the many factors involved, determining the scale of port logistics parks is a complex task. In the context of the current scarcity of land resources, reasonably determining the scale of the port logistics park reasonably is very necessary. This paper combines three models—the analogy model, land use intensity model, and space–time consumption model—into a new model. These three models have their own shortcomings, which we aim to avoid through their combination. The new model is applied to calculate the scale of a port logistics park, with Xiamen Haicang Port as a case for actual verification. In 2035, the container throughput of Xiamen Haicang Port will reach 14 million TEU, and the required scale of the port logistics park is 3.2 km^2. The scale of the port logistics park aligns with the planning requirements. Practice has proven that this method is feasible for calculating the scale of logistics parks.

Keywords Logistics park · Combined model · Land area

S. Sun · P. Cai (✉) · J. Fu
Center for Economics, Policy and Development Strategy, Waterborne Transport Institute, Beijing, China
e-mail: caipeng@wti.ac.cn

S. Sun
e-mail: sunsw@wti.ac.cn

J. Fu
e-mail: fujinxiu@wti.ac.cn

K. Subramaniyam and M. Paprzycki (eds.), *Smart Cities and the Internet of Things*, Internet of Things, https://doi.org/10.1007/978-3-032-08290-9_6

1 Introduction

In the global supply chain, the role of ports has expanded from a single cargo transfer center to providing a full range of value-added services, including warehousing, distribution, processing and modification, and packaging. Ports have formed a powerful modern logistics system and become a hub for the flow of goods, capital, and information. Logistics parks are the inevitable product of the development of the modern logistics industry to a certain stage. The purpose of establishing logistics parks is to achieve the intensification of logistics facilities, the commonality of logistics operations, and the rationalization of the spatial layout of urban logistics facilities. The concept of a logistics park has also been mentioned in other countries. The scale and layout of logistics park land directly impact port operations.

Logistics parks have been discussed in many literatures [1]. Port logistics parks refer to the centralized construction of logistics professional facilities and many professional logistics enterprises in the region near and relying on the port, which is a physical gathering place. Port logistics parks are logistics economic regions with a certain scale and comprehensive service functions.

Reasonable planning and effective use of land in logistics parks can improve the operational efficiency of ports, reduce logistics costs, and enhance the competitiveness of ports. The optimal allocation of logistics park land can promote the coordinated development of the port and the hinterland economy, form a positive interaction, and promote regional economic growth. The supply of logistics park land directly affects the development space of the port. Insufficient supply or unreasonable allocation of logistics land may lead to rising logistics costs and affect the operational efficiency and service quality of the port. The container throughput of Shanghai Waigaoqiao Port is about 19 million twenty-foot equivalent units (TEU). To save resources, Shanghai Waigaoqiao Port reduced the scale of the logistics park land from 3.83 k^2 to 3.14 k^2 (as shown in Fig. 1). However, determining the scale of port logistics land is not one-size-fits-all; this remains a problem to be solved.

Based on the above background, this paper constructs a new land use calculation model for port logistics parks [2]. Operations research theory is also applied in logistics park planning research [3]. We select three types of port logistics land calculation models and methods: analogy model, land use intensity model, and space–time

Fig. 1 Port logistics land area (km^2)

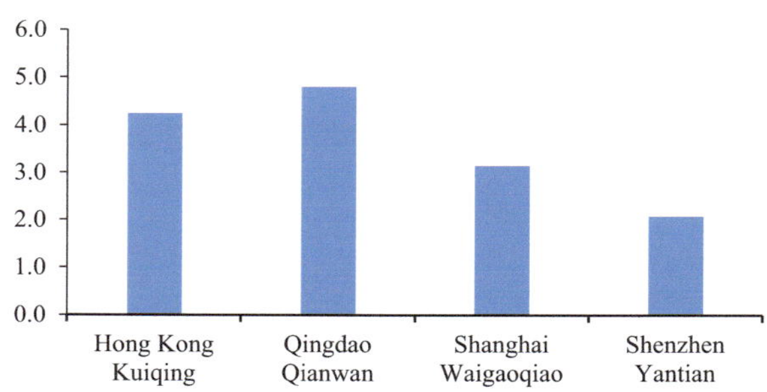

consumption model [4]. We derive a new calculation model through the combination of these models, striving for more accurate results. In the case study, this paper takes the future logistics land scale of Xiamen Haicang port as the measurement object. We can determine the reasonable scale of port logistics land through calculation. The research results of this paper can ensure the adequate supply and reasonable layout of logistics land, thereby promoting the high-quality development of ports and avoiding the waste of land resources.

2 Model Construction

2.1 Model Introduction

Many have proposed models for the size and location of seaports through the theory of operational research and integer programming. There are also many researchers in China using bi-level programming models to solve the scale of logistics centers. The three port logistics land scale calculation models applied in this paper are the analogy model, land use intensity model, and space–time consumption model.

- Analogy model: It is necessary to refer to the scale of other similar port logistics parks to determine the scale of each logistics park in the planning area. In the calculation, the economic development level and logistics demand of the target area must be taken into consideration.
- Land use intensity model: According to the land use intensity coefficient specified in relevant national standards, we calculate the scale of logistics land based on three indicators: the annual container throughput of the target port, the proportion of containers entering the logistics park, and the logistics intensity of the logistics park.
- Space-time consumption model: Goods require a certain amount of time and space resources when they are reloaded or stored in the logistics park. We establish a mathematical model by balancing the space-time resources demanded by the cargo and those provided by the logistics park.

2.2 Analogy Model

The analogy model is as follows:

$$A_1 = \mathrm{G}/P \tag{1}$$

A_1 Analogy model port logistics land scale.

G Logistics park cargo turnover; in this paper, it refers to port throughput.

P Logistics land allocation coefficient. The calculation of this index is based on cases of existing typical ports, including Shenzhen Yantian Port, Shenzhen Dachan Bay Port, Qingdao Qianwan Port, Shanghai Waigaoqiao Port, Hong Kong Kwai Tsing Port, and Ningbo Meishan Port.

2.3 Land Use Intensity Model

The land use intensity model is as follows:

$$A_2 = (G \times L)/N \tag{2}$$

A_2 Analogy model port logistics land scale.
G Logistics park cargo turnover; in this paper, it refers to port throughput.
L Average proportion of goods entering the logistics park. This index is related to the mode of collection and distribution. The goods are less transported by railway and waterway into the park. According to survey data, the proportion of goods transported by railway and waterway is 10%, while the proportion of goods transported by road is 40%.
N Park logistics intensity. This indicator represents the amount of logistics activity that occurs per unit area of land. It is usually used to measure the intensity of logistics activities and land use efficiency within the park. This index helps evaluate the logistics operation efficiency of the park and the utilization of land resources.

2.4 Space–Time Consumption Model

$$A_3 = \alpha \sum_{i=1}^{n} (V_i T_i Q_i F_i / 365 S_i) \tag{3}$$

A_3 Space–time consumption model [5] of port logistics land scale.
n Types of goods in the logistics park.
α Proportion of the total area of the logistics park and warehouse area.
V_i Average volume of unit weight of category i goods.
T_i Average turnaround time (the time of the goods in the warehouse of category i goods).
Q_i Annual turnover of category i goods.
F_i Time non uniformity coefficient of category i goods.
S_i Volume of category i goods that can be stored per square meter of warehouse area.

2.5 A New Combined Model

In the process of determining the area of logistics land, the three models have their own shortcomings. The disadvantage of the analogy model is that the natural resource conditions and economic and social environment of the area where the logistics land is available for reference are different. The disadvantage of the land use intensity model is that the determination of the logistics intensity index of the park has a strong subjective element. The disadvantage of the space–time consumption model is that there are many parameters needed, and the data is difficult to obtain. It is necessary to avoid the shortcomings of each model through the establishment of a combined model to obtain relatively accurate results. The scale of the port logistics park is synthesized from the calculation results of the analogy model, the land use intensity model, and the space–time consumption model.

$$A = x_1A_1 + x_2A_2 + x_3A_3 \tag{4}$$

x_1,x_2, x_3: index weigh, $x_1 + x_2 + x_3 = 1$.

3 Case and Calculation

3.1 The Basic Situation of Xiamen Haicang Port

This paper examines the scale calculation of the Xiamen Haicang Port Logistics Park as a case study. With a container throughput of exceeding 12 million TEU, Xiamen Port ranks seventh among China's coastal ports, following Shanghai Port, Ningbo Zhoushan Port, Shenzhen Port, Qingdao Port, Guangzhou Port, and Tianjin Port (as shown in Fig. 2). Haicang Port is the most concentrated area of containers in Xiamen. The Haicang Port container throughput was 9.11 million TEU in 2022 and will reach 14 million TEU by 2035 [6]. The length of the port shoreline is 6332 m, which will be increased to 7148 m by 2035.

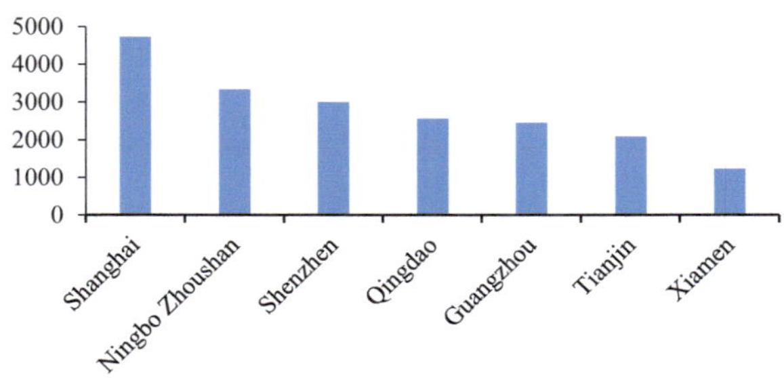

Fig. 2 Container throughput of major ports

Table 1 Logistics matching coefficient of the park

Port	Coefficient
Yantian port	702
Dachan bay port	545
Qianwan port	535
Waigaoqiao port	493
Kuiqing wharf	471
Meishan port	362

3.2 Analogy Model Results

We refer to the cases of Yantian Port of Shenzhen, Dachan Bay Port of Shenzhen, Qianwan Port of Qingdao, Waigaoqiao Port of Shanghai, Kaiqing Wharf of Hong Kong, and Meishan Port of Ningbo. Through these cases, it has been established that a port logistics park logistics support coefficient within the range of 4–6 million TEU/km^2 is more reasonable (as shown in Table 1). The average logistics matching coefficient for the sample port logistics park is 5 million TEU/km^2.

In this paper, we select a port logistics park logistics support coefficient of 5 million TEU/km^2. By 2035, the container throughput of Haicang Port is expected to reach 14 million TEU. According to the analogy model, the scale of Xiamen Haicang Port Logistics Park is calculated to be 2.8 km^2.

$$A_1 = \frac{G}{P} = \frac{1400}{500} = 2.8 \tag{5}$$

3.3 Land Use Intensity Model Results

The average proportion of goods entering the logistics park is considered. By 2035, the container throughput of Xiamen Haicang Port is expected to reach 14 million TEU. The proportion of container throughput completed by different collection and distribution modes varies. In 2035, the proportion of container transportation in Xiamen Haicang Port is projected to be 40% by water, 10% by rail, and 50% by road (as shown in Fig. 3). Based on these proportions, the average proportion of goods entering the logistics park is 25% ($50\% \times 10\% + 50\% \times 40\% = 25\%$).

$$L = 25\% \tag{6}$$

China's national standards require that the logistics intensity of logistics parks in coastal ports should exceed 10 million tons/km^2 per year (γ). According to the enterprise survey data, the highest logistics intensity of China's logistics parks is 15

Fig. 3 The proportion of the container transportation mode in Xiamen Haicang port

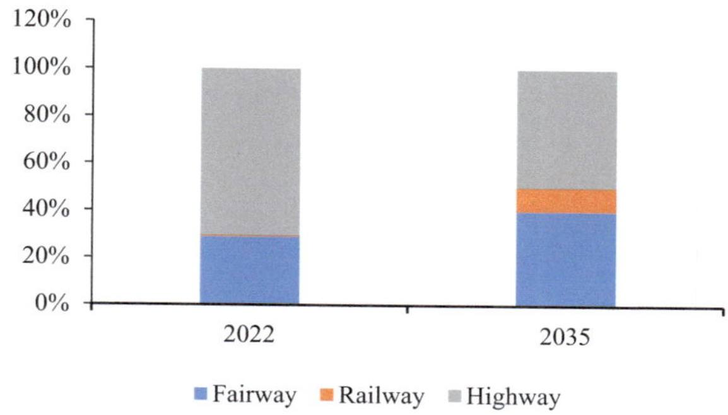

million tons/km^2 per year (δ). The average weight of a container in Xiamen Port is 12.5 tons (ε). Based on these parameters, the logistics intensity of Xiamen Haicang Port Park is 1 million TEU/km^2 per year.

$$N = \frac{\gamma + \delta}{2} \times (1/12.5)$$

$$N = 100 \tag{7}$$

According to the land use intensity model, the scale of Xiamen Haicang Port Logistics Park is approximately 3.5 km^2.

$$A_2 = (G \times L)/N$$

$$A_2 = 3.5 \tag{8}$$

3.4 *Space–Time Consumption Model Results*

According to an actual investigation, the goods entering Haicang Port Logistics Park are primarily containerized. The average turnover time of goods is 18 days.

According to the space–time consumption model, the land use scale of Xiamen Haicang Port Logistics Park is 3.3 km^2.

$$A_3 = \alpha \sum_{i=1}^{n} (V_i T_i Q_i F_i / 365 S_i)$$

$\alpha = 0.5$	50% of Haicang Port's area is designated as warehouse space.
$V = 1.34$ m^3/t	The average volume of a container per unit weight is 1.34 m^3/t.

Table 2 The results table

	Weigh	Scale (km^2)
Analogy model results (A_1)	1/3	2.8
Land use intensity model results (A_2)	1/3	3.5
Space–time consumption model results (A_3)	1/3	3.3
The scale of the logistics park	3.2	

$T = 8$ The average turnaround time for a container is 8 days.
$Q = 17{,}500$ Haicang Port's container throughput is 175 million tons.
$F = 1$ The time correlation coefficient of the container is 1.
$S = 3\ m^3/m^2$ The number of containers that can be stored per square meter of warehouse area is 3 m^3/m^2.

$$A_3 = 3.3 \tag{9}$$

.

3.5 *Final Test Result*

We analyzed the results from the analogy model, the land use intensity model, and the space–time consumption model. Applying Formula (4) for calculation, the scale of the Xiamen Haicang Port Logistics Park is determined to be 3.2 km^2 (as shown in Table 2). The calculation results align with the requirements of the Xiamen Freight Transportation System Planning (2020–2035)[7].

4 Conclusion

In this paper, the three models—analogy model, land use intensity model, and space–time consumption model—are combined into a new model to mitigate the shortcomings of each individual model. The new model is used to measure the scale of the Xiamen Haicang Port Logistics Park in 2035. After the calculation, it is determined that the scale of Xiamen Haicang Port Logistics Park in 2035 should be set at 3.2 km^2. The new model overcomes the limitations of the traditional three models, and the calculation conclusions are also utilized in Xiamen's cargo transportation planning. The calculation method presented in this paper is generalizable and holds reference significance for calculating the scale of logistics parks in other ports. However, the parameter acquisition for the model in this paper is more complex and involves certain subjective factors. Moreover, port logistics parks differ from other logistics parks. The model applicable to all types of logistics parks needs to be studied.

Acknowledgements The research results of this paper primarily stem from key projects of the Ministry of Transport of the People's Republic of China, with some data obtained through field research.

References

1. Ruske, W.: City Logistics-Solutions for Urban Commercial Transport by Cooperative Operation Management. OECD Seminar on Advanced Road Transport Technologies (1994)
2. Zhang, Z., Ge, Z.: Comparative study on prediction methods of logistics land scale. Logist. Technol. **36**(5) (2013)
3. Drezner, Z.: Facility Location-A Survey of Applications and Methods. Springer, Berlin (1995)
4. Ding, J.: Research on spatial layout optimization of logistics industry in Qingdao West Coast New Area. Shandong University of Finance and Economics (2022)
5. Bai, T., Ma, Y., Sun, Z.: Evaluation of urban rail transit carrying capacity based on space-time consumption model. Compr. Transp. **45**(10) (2023)
6. Chen, L.: Xiamen port cargo throughput forecast and port logistics development countermeasures. Railw. Procure. Logist. (12) (2022)
7. Wang, W.: Xiamen port container terminal rear land space demand forecast. Water Transp. Manag. **45**(7) (2023)

Intelligent Connected Vehicles Actively Coordinate Collision-Avoidance Control Based on Lane-Change Trajectory Prediction of Sidecars

Dengcheng Liu, Yifan Hu, Zhipeng Sun, Zhouping Zhang, Ruiyang Xiong, and Yunfeng Chen

Abstract The autonomous selection of obstacle avoidance speed and power source torque distribution mode for intelligent connected vehicles is affected by the original lane-changing trajectory of the side vehicle, which limits the initiative of collision avoidance and has a low safety factor. Therefore, a prediction model for lane-changing trajectory of the side vehicle based on a long short-term memory network is proposed. A bidirectional long short-term memory network was applied to predict the lane-change trajectory of the side vehicle, and a feedforward and feedback control structure was introduced to control the deviation between the vehicle and the expected path. Controlling the directional deviation $\Delta\theta$ at the vehicle center of mass point and the lateral deviation at the preview point, we endow the obstacle avoidance process with active attributes, and coordinate the control of the collision avoidance path of intelligent connected vehicles. The simulation experiment results show that the sidecar lane-change trajectory prediction error is relatively low at different vehicle speeds, especially at medium and high speeds, with significant advantages. Compared with other methods, the prediction accuracy is higher and the error growth is slower. In addition, comparing the obstacle avoidance effects of three obstacle avoidance methods by setting obstacle points on the designated lane, the proposed method has the shortest obstacle avoidance path distance, no unnecessary obstacle avoidance actions, and a high safety factor, indicating that it can effectively achieve active coordinated collision avoidance control for intelligent connected vehicles and improve driving safety and adaptability.

Keywords Lane-change trajectory prediction · Intelligent networking · Collision avoidance control · Feedback control

---

D. Liu (✉) · Z. Zhang · R. Xiong
Nanchang Automotive Institute of Intelligence and New Energy, Nanchang, Jiangxi, China
e-mail: liudengcheng0791@163.com

Y. Hu · Y. Chen
Nanchang TongYan New Material Technology Co., Ltd., Nanchang, Jiangxi, China

Z. Sun
School of Automotive Engineering, Hangzhou Polytechnic, Hangzhou, Zhejiang, China

K. Subramaniyam and M. Paprzycki (eds.), *Smart Cities and the Internet of Things*, Internet of Things, https://doi.org/10.1007/978-3-032-08290-9_7

1 Introduction

With the rapid development of communication technology, the advantages of vehicle networking technology in intelligent driving environment recognition are becoming increasingly apparent, providing important support for active vehicle safety [1]. However, traditional collision avoidance methods mostly rely on vehicle-to-vehicle communication to obtain information and establish collision avoidance models, which have shortcomings such as relying on the installation of various sensors to achieve collision avoidance functions, difficulty in forming a unified standard, and limited effectiveness in vehicle queue driving control [2, 3]. Therefore, the research on active coordination and collision avoidance control methods for intelligent connected vehicles is particularly important. It cannot only effectively improve the safety of intelligent vehicle driving but also promote the development of intelligent connected vehicles, achieve the coordinated operation of elements in the four fields of "human vehicle road environment," and make transportation safer, more convenient, and energy-saving.

Currently, scholars are actively exploring the use of various technologies for obstacle recognition and path planning. Among them, in [4], scholars proposed an active collision avoidance method for autonomous vehicles based on a single-parent genetic algorithm. However, this method relies on the environmental perception system of autonomous vehicles, and if the perception system malfunctions or errors occur, it may affect the planning and execution of collision avoidance paths. In [5], an autonomous driving collision avoidance method based on drivable areas was proposed. However, in extreme environments such as extreme weather or complex terrain, the collision avoidance effect may be affected. In [6], researchers proposed an adaptive collision avoidance control based on the extensible decision method, which designed a sliding mode observer to estimate the longitudinal force of the tire. However, the forgetting factor and parameters of the recursive least squares method need to be adjusted and optimized according to specific situations, which increases the difficulty and complexity of the method's application.

Although existing car collision avoidance control methods have made some progress in obstacle recognition and path planning, these methods still face shortcomings such as strong dependence on environmental perception systems, limited adaptability in extreme environments, and complex parameter adjustments. To address the above shortcomings, this paper studies the active coordination and collision avoidance control methods of intelligent connected vehicles under the prediction of lane-changing trajectories of side vehicles. This method innovatively utilizes bidirectional long short-term memory (Bi-LSTM) networks to accurately predict the lane-changing trajectory behavior of surrounding vehicles, solving the problem of the strong dependence of autonomous vehicles on environmental perception systems. Based on a feedforward feedback control structure, it responds in real time to extreme weather or complex terrain changes, ensuring obstacle avoidance control effectiveness. The parameter adjustment is relatively simple, enhancing driving safety and adaptability in complex traffic environments.

2 Intelligent Connected Vehicles Actively Coordinate Collision Avoidance Control

2.1 Prediction of Lane-Change Trajectory for Sidecars

The forget gate mainly determines which information elements contained in the latent features of the target vehicle state observation feature vector should be removed from the memory cells and forgotten; the input gate is mainly responsible for performing effective control operations on the input of memory cells in the network, based on the valid data output in the previous moment and the valid data input in the current moment; and the main responsibility of memory cells is to perform effective storage operations on the state vectors generated by the current situation of the cell [7].

In practical work, a convolutional neural network uses the sigmoid function to extract potential features from the target vehicle state observation feature vector and inputs them into a bidirectional long short-term memory network. Through forward and backward LSTM layers, effective flow control and capture operations are performed on the potential features of the target vehicle state observation feature vector. Finally, the corresponding predicted lane-change trajectory of the side vehicle can be output, which is:

$$y_t = w_f + w_b + b_y \tag{1}$$

Among them, y_t represents the predicted lane-change trajectory of the sidecar output by the bidirectional long short-term memory network; w_f represents the effective weight between the forward LSTM layer and the output layer; w_b represents the effective weight between the backward LSTM layer and the input layer; and b_y represents the effective bias of the latent features of the target vehicle state observation feature vector in the hidden units of the bidirectional long short-term memory network.

To optimize the lane-change trajectory prediction model based on a bidirectional long short-term memory network, an attention mechanism is introduced to improve the prediction accuracy.

If the attention weight is set to α_t, the final prediction result of the lane-change trajectory of the sidecar can be expressed as:

$$Y = \sum_{t=1}^{T} \alpha_t h_t \tag{2}$$

Among them, h_t is the hidden state of the bidirectional long short-term memory network at time step t, and T is the total number of time steps.

The time complexity calculation of the side vehicle lane-changing trajectory prediction algorithm based on a long short-term memory network and attention mechanism is as follows:

$$O = T(dh + h^2 + qk) \tag{3}$$

In the formula, T is the length of the input sequence, the number of hidden layer units is h, the input feature dimension is d, and the dimensions of the attention mechanism key and value are q and k, respectively.

The spatial complexity calculation of the side vehicle lane-changing trajectory prediction algorithm based on a long short-term memory network and attention mechanism is as follows:

$$P = dh + h + Th + d(q + k + v) + Tv \tag{4}$$

In the formula, v is the feature weight.

LSTM and attention mechanisms can effectively capture long-term dependencies and important information in time series data and can reduce time and space complexity through data preprocessing.

2.2 *Implementation of Active Coordinated Collision Avoidance Control for Cars*

To achieve active coordinated collision avoidance control for automobiles, this paper introduces a feedforward and feedback control structure based on the predicted lane-change trajectory of the side vehicle. Adopting feedforward control to overcome road curvature interference and control the deviation between the vehicle and the expected path.

Combining the state formula (3) and the objective function (4), the optimal feedback control law is expressed as:

$$\delta_{fb} = -K \cdot x(k) \tag{5}$$

Finally, the steering wheel control input angle is the sum of the feedforward and feedback parts:

$$\delta = \delta_{ff} + \delta_{fb} \tag{6}$$

Based on the predicted lane-change trajectory of the sidecar, a feedforward and feedback control structure is introduced to overcome road curvature interference and system uncertainty. Feedforward control utilizes known input information, such as road curvature and vehicle speed, combined with system dynamic characteristics to calculate the required control inputs, such as steering wheel angle, to compensate for external disturbances in advance. At the same time, the feedback controller takes directional deviation and lateral deviation as control variables and adopts the linear quadratic regulator (LQR) algorithm design to adjust the vehicle's driving state in real time, ensuring that the vehicle travels along the expected trajectory and ultimately achieves active coordinated collision avoidance control of the car.

Considering that controlling vehicles requires the cooperation of other intelligent connected vehicles to achieve collision avoidance, a collaborative collision avoidance strategy for connected vehicles is designed: intelligent connected vehicles exchange information directly through 5G wireless communication technology. This communication method has the characteristics of low latency and high reliability and can transmit key information such as the vehicle's position, speed, acceleration, and driving intention in real time. At the same time, the vehicle communicates with the roadside unit (RSU) to upload its information to the infrastructure platform, which then broadcasts the relevant information to surrounding vehicles. The vehicle information transmitted and shared includes the current location (longitude and latitude), speed, acceleration, heading angle, lane-changing intention, braking intention, road conditions ahead (such as obstacles or construction areas), traffic signal status, etc.

Then we design the distributed collaborative collision avoidance decision-making based on wireless communication technology. When the controlling vehicle detects potential collision hazards and plans to change lanes for collision avoidance, it sends negotiation requests to surrounding vehicles through communication mechanisms, including its driving intention and expected lane-change trajectory. After receiving the request, other vehicles evaluate whether they can cooperate with the lane-change operation of the vehicle based on their own driving status and environmental information. If they can cooperate, they will send confirmation messages to the requesting vehicle and adjust their driving status (such as slowing down and maintaining distance); if unable to cooperate, they will send a rejection message to the requesting vehicle and provide their suggestions. They request the vehicle to readjust its lane-changing strategy or find other collision avoidance solutions based on the feedback received to achieve vehicle networking collaborative collision avoidance.

In the process of collaborative collision avoidance, the feedforward control strategy designed in this paper predicts potential collision hazards in advance based on communication information between vehicles and pre-set rules and takes corresponding preventive measures. When the vehicle receives the lane-change intention information from the vehicle in front, it can adjust its speed or maintain a safe distance in advance according to its speed and position to avoid a collision. Feedback control adjusts and modifies the results of feedforward control based on real-time vehicle status and environmental information. During the process of changing lanes, if it is found that the actual driving status of other vehicles does not match expectations, feedback control should be used to adjust one's own driving strategy in a timely manner to ensure the success of collision avoidance.

3 Simulation Experiment Research

In the experiment, the vehicle length was 4.2 m, the vehicle width was 1.8 m, and the lane width was 3.75 m. We select multiple different vehicle speeds as experimental conditions, namely low speed (30 km/h), medium speed (60 km/h), and high speed (90 km/h). At each vehicle speed, multiple lane-changing experiments are conducted, equipped with a high-precision Global Positioning System/Inertial Navigation System (GPS + INS) fusion positioning system for real-time recording of the vehicle's driving trajectory, speed, acceleration, and other information. The technical parameters are shown in Fig. 1.

The format, encoding, and semantics of data should be consistent. Statistical analysis was conducted on the preprocessed data to calculate the prediction performance of the single-parent genetic algorithm, the extensible decision-making method, and the proposed method for the lane-change trajectory of side vehicles. The results are shown in Figs. 1, 2, and 3.

From Figs. 1, 2, and 3, it can be seen that when the vehicle speed is low (30 km/h), the prediction error of the lane-change trajectory of the three methods is relatively low, and the differences among the three methods are minimal.

When the vehicle speed is medium (60 km/h), the prediction errors of the sidecar lane-change trajectory of the three methods gradually widen. Among them, the extensible decision method has the highest prediction error of the sidecar lane-change trajectory, followed by the single-parent genetic algorithm, and the proposed method has the smallest prediction error of the sidecar lane-change trajectory.

When the vehicle speed is high (90 km/h), there is a greater difference in the prediction error of the sidecar lane-change trajectory among the three methods. This is because the increase in vehicle speed increases the computational complexity and difficulty of switching to trajectory prediction. If the prediction method relies on the environmental perception system and cannot adapt to the real-time complex changes of the lane, it will increase the prediction error.

Fig. 1 Prediction results of lane-change trajectory for vehicles at low speed (30 km/h)

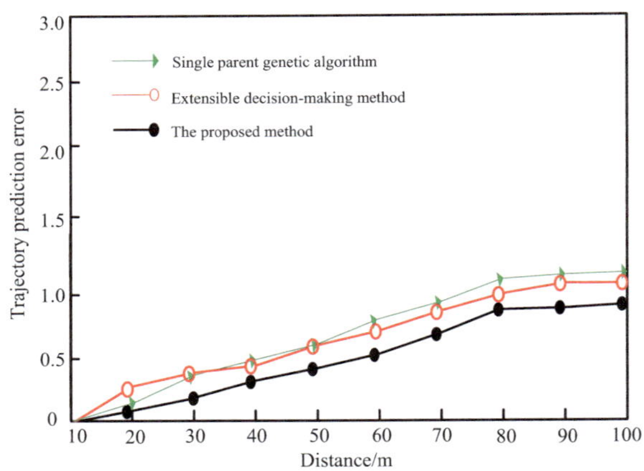

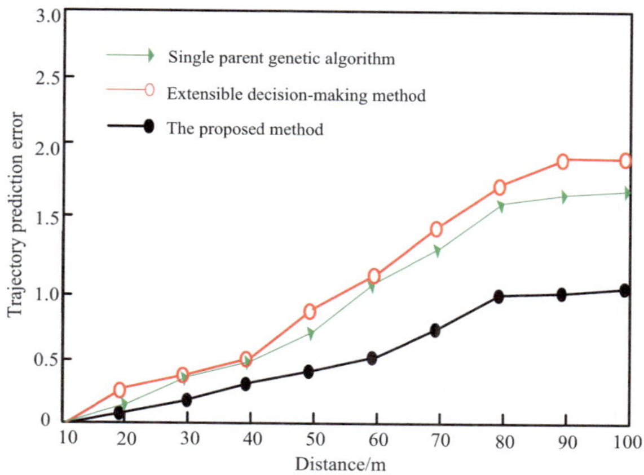

Fig. 2 Prediction results of lane-change trajectory for the lower side vehicle at medium speed (60 km/h)

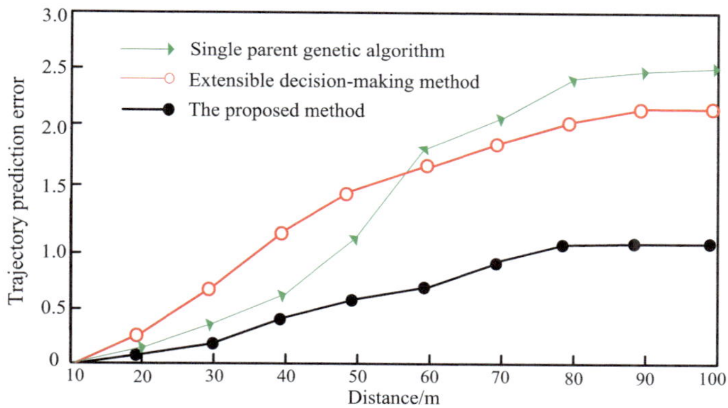

Fig. 3 Prediction results of lane-change trajectory for vehicles under high-speed (90 km/h)

At this time, the advantage of the proposed method is more obvious, with a prediction error of <1.2 for the sidecar lane-change trajectory. This indicates that when predicting the sidecar lane-change trajectory at different vehicle speeds, the proposed method exhibits lower prediction errors at low, medium, and high speeds, especially at medium and high speeds. Compared with the extensible decision method and the single-parent genetic algorithm, its prediction accuracy is higher, and the error growth is slower. This is because the proposed method integrates long short-term memory (LSTM) networks and attention mechanisms to establish a prediction model for the lane-changing trajectory of side vehicles, which can adapt to real-time complex lane changes in a timely manner, thus achieving accurate prediction of the lane-changing trajectory of side vehicles. However, scalable decision-making methods and single-parent genetic algorithms rely on environmental perception data and cannot respond to lane changes in a timely manner, resulting in significant errors.

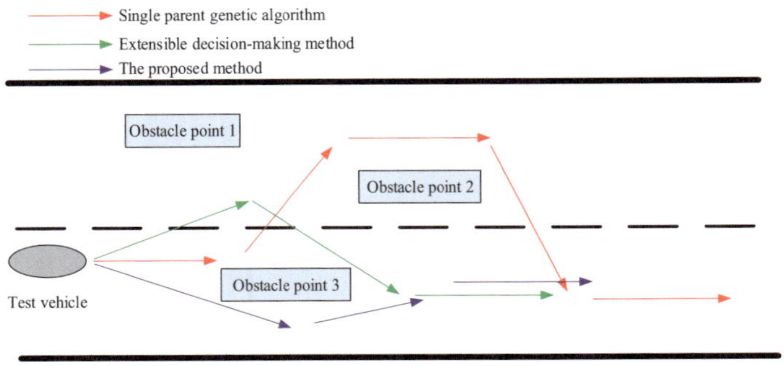

Fig. 4 Obstacle avoidance path planning

Further setting up three obstacle points on the designated lane and comparing the obstacle avoidance effects of the three methods, the results are shown in Fig. 4.

According to Fig. 4, the obstacle avoidance path planned by the single-parent genetic algorithm is the longest and contains unnecessary actions; the obstacle avoidance path distance planned by the extensible decision-making method is relatively reasonable, but there are also unnecessary obstacle avoidance actions, and the safety factor of the obstacle avoidance is relatively low; and the obstacle avoidance path planned by the proposed method has a short distance, no unnecessary obstacle avoidance actions, and a high safety factor for the entire obstacle avoidance route. Based on comparative analysis, it can be determined that the proposed method has the best obstacle avoidance effect, indicating that it can effectively achieve active coordination and collision avoidance control of intelligent connected vehicles.

4 Conclusion

This article proposes an intelligent networked vehicle active coordination collision avoidance control method for predicting lane-changing trajectories of side vehicles. Long short-term memory networks can capture long-term dependencies in time series data, which is crucial for predicting vehicle driving trajectories. However, relying solely on LSTM is still difficult to fully cope with the ever-changing road environment. Therefore, this study introduces an attention mechanism that can dynamically adjust the weights of different time step information, making the model more focused on the key information that has the greatest impact on the prediction results. At the same time, a composite control structure of feedforward and feedback is adopted, which fully considers the various interference factors faced by intelligent connected vehicles in actual operation.

References

1. Chen, P., Kong, L., Peng, Y.: Obstacle avoidance path planning for unmanned vehicle based on fifth-order Bezier curve. Sci. Technol. Eng. **22**(31), 13933–13941 (2022)
2. Gim, S., Lee, S., Adouane, L.: Safe and efficient lane change maneuver for obstacle avoidance inspired from human driving pattern. IEEE Trans. Intell. Transp. Syst. **23**(03), 2155–2169 (2022)
3. Dong, Z., Yan, L.U., Shi, G.T.S.W.: WatchDog: real-time vehicle tracking on geo-distributed edge nodes. ACM Trans. Internet Things **4**(1), 1–23 (2023)
4. Tian, G., Dai, P.: Active collision avoidance method of driverless vehicle based on partheno genetic algorithm. J. Jilin Univ. **53**(08), 2404–2409 (2023)
5. Wu, P., Zhang, W., Wu, J., Cheng, J.: Research on an autonomous driving collision avoidance method based on drivable area. Intell. Comput. Appl. **13**(12), 107–113 (2023)
6. Ni, L., Bai, Q., Zhang, Z.: Abnormal fault detection of new energy vehicle motor based on AdaBoost algorithm. Comput. Simul. **41**(4), 97–101 (2024)
7. Echab, H., Khallouk, A., Ez-Zahraouy, H.: Effect of connected and autonomous vehicles on traffic flow at a bidirectional road. Int. J. Mod. Phys. C **33**(07), 1–10 (2022)

A CRITIC-Cloud Model Coupling Framework for Dynamic Evaluation and Optimization of Urban Rail Transit Emergency Plans

Chaochen Wang, Wenkai Xu, Jun He, Xinhuan Zhang, and Jinhong Wu

Abstract Urban rail transit emergency plans are critical for managing operational disruptions, yet existing evaluation methods often suffer from subjective biases and non-quantifiable results. This study proposes a novel evaluation framework integrating the CRITIC method and cloud model theory to address these limitations. The CRITIC method objectively assigns weights to evaluation indicators by analyzing inter-criteria correlations and conflicts, while the cloud model translates qualitative assessments into quantitative representations using numerical features. An evaluation indicator system is established, encompassing comprehensiveness, operability, and cost-effectiveness. Case study results demonstrate that the comprehensive evaluation cloud aligns the emergency plan between "Average" and "Good," highlighting its practical applicability despite areas for improvement. This method effectively reduces subjectivity, enhances visualization of results through cloud mapping, and provides a robust theoretical foundation for optimizing urban rail transit emergency preparedness.

Keywords Urban rail transit · Emergency plan evaluation · Cloud model · CRITIC method · Quantitative assessment

C. Wang · W. Xu · X. Zhang · J. Wu
School of Engineering, Zhejiang Normal University, Jinhua, China
e-mail: 05118176@zjnu.edu.cn

W. Xu
e-mail: xuwenkai@zjnu.edu.cn

X. Zhang
e-mail: zxh@zjnu.cn

J. Wu
e-mail: jhwjh@zjnu.cn

J. He (✉)
School of Automotive, Yuhuan Secondary Vocational and Technical School, Yuhuan, China
e-mail: 1564381696@qq.com

K. Subramaniyam and M. Paprzycki (eds.), *Smart Cities and the Internet of Things*, Internet of Things, https://doi.org/10.1007/978-3-032-08290-9_8

1 Introduction

As a high-capacity, efficient, and environmentally friendly passenger transportation system, urban rail transit plays a vital role in meeting commuter demands, alleviating urban congestion, and promoting sustainable socioeconomic development. Given the characteristic high passenger volumes and unique operational environments of urban rail systems, any operational disruption may readily escalate into adverse consequences, including property damage and casualties. Functioning as strategic response frameworks, urban rail emergency plans provide concrete operational guidance for disaster mitigation efforts. Therefore, ensuring the scientific rationality of emergency plans becomes imperative to enable coordinated multi-agency responses during contingencies, thereby facilitating timely incident resolution and minimizing human casualties and economic losses.

Subsequent studies have investigated emergency plan quality assessment methodologies. Ju et al. [1] constructed an evaluation index system for emergency response capacity using fuzzy Analytic Hierarchy Process (AHP), and combined the 2-tuple fuzzy linguistic approach to score sub-criteria and calculate the overall emergency response capacity of alternative plans, providing a scientific basis for emergency decision-making. Zhang et al. [2] constructed an evaluation framework using the technique for order preference by similarity to an ideal solution–grey relational analysis (TOPSIS–GRA) method, integrating subjective and objective weights to quantitatively assess and rank the emergency evacuation capacity of eight Beijing metro stations, and validated the robustness and accuracy of the method. Luu et al. [3] utilized the TOPSIS method combined with multiple linear regression to analyze Vietnam's national disaster database, assess flood risk across regions and provinces nationwide, and create a flood risk map. The cloud model [4], however, systematically integrates these uncertainties through qualitative–quantitative mapping. This model characterizes qualitative concepts through three numerical features (E_x, E_n, H_e) rather than explicit functions, effectively combining fuzziness and randomness. The application of cloud modeling enhances evaluation accuracy by mitigating data distortion during qualitative–quantitative conversions, enabling flexible variable representation while reducing subjective biases. The model's bidirectional conversion capability between qualitative and quantitative metrics resolves stochastic evaluation issues that could compromise scientific objectivity. Resilient to interference from fuzziness and randomness, the cloud model effectively processes uncertain and ambiguous evaluation indicators [5]. Consequently, the proposed cloud model-based evaluation framework for urban rail emergency plans addresses event uncertainties while providing novel methodological insights for emergency preparedness research in transit systems.

The remainder of this paper is structured as follows: Sect. 2 introduces the methods, Sect. 3 describes the case study and presents its results and Sect. 4 presents the conclusions.

2 Method

2.1 The CRITIC Method

Conventional weight determination methodologies in emergency plan evaluation, including the entropy method [6], analytic hierarchy process (AHP) [7], and grey relational analysis [8], share a fundamental limitation: their inability to account for inter-indicator correlations and inconsistencies. The Criteria Importance Through Intercriteria Correlation (CRITIC) method, developed by Joosep et al. [9], uses objective weighting through inter-indicator correlation analysis. This approach determines weights by quantifying comparative advantages and conflicts between indicators, eliminating evaluator bias, and aligning with the emergency plans' dynamic and comprehensive nature. Therefore, the CRITIC method was applied to the evaluation of the urban rail transit emergency plan to assign optimal weights across indicator categories [10].

The weight determination procedure through the CRITIC methodology proceeds as follows [11].

Step 1: Construct the indicator data matrix. For p evaluators and m evaluation indicators in urban rail transit emergency plan assessment, the initial data matrix R is formulated as follows:

$$R = \begin{bmatrix} r_{11} & \cdots & r_{1j} & \cdots & r_{1m} \\ \vdots & \ddots & \vdots & \vdots & \vdots \\ r_{i1} & \cdots & r_{ij} & \cdots & r_{im} \\ \vdots & \ddots & \vdots & \vdots & \vdots \\ r_{p1} & \cdots & r_{pj} & \cdots & r_{pm} \end{bmatrix}, i = 1,2, \ldots, p, j = 1,2, \ldots, m. \quad (1)$$

Step 2: Calculate the standard deviation σ_j for each j indicator, representing comparative intensity.

Step 3: Determine conflict coefficient y_j between the j indicator and others:

$$\mathbf{y} = \sum_{l=1}^{m}\left(1 - \frac{\mathrm{Cov}(j, l)}{\sigma_j \sigma_l}\right), j = 1,2, \ldots m. \quad (2)$$

where $\mathrm{Cov}(j, l)$ denotes the covariance between indicators j and l, and σl represents the standard deviation of the l indicator.

Step 4: Compute the comprehensive information measure C_j.

$$C_j = \sigma_j \gamma_j, j = 1,2, \ldots, m. \quad (3)$$

Step 5: Derive final indicator weights W_j.

$$W_j = \frac{c_j}{\sum_{j=1}^{m} c_j}, j = 1,2, \ldots, m. \tag{4}$$

2.2 Cloud Model

2.2.1 Fundamental Concepts and Numerical Characteristics of Cloud Model

The cloud model theory, pioneered by Academician Li et al. [4] in the 1990s, establishes a mathematical framework for uncertainty transformation between qualitative natural language descriptions and quantitative representations. Recent years have witnessed expanding applications of cloud model theory in rail transit domains, particularly in passenger flow forecasting, safety assurance, reliability analysis, and risk assessment [12].

Formally, let U denote a quantitative universe discourse with precise numerical representation, and B be a qualitative concept defined on U. For any $x \in U$, as a stochastic instantiation of concept B, the certainty degree $A(x) \in [0, 1]$ satisfies:

$$A : U \rightarrow [0,1], \forall x \in U, x \rightarrow A(x) \tag{5}$$

The distribution of x across U constitutes a cloud, with each x termed a cloud droplet.

The cloud's numerical characteristics [5] comprise: Expectation (E_x)—the central tendency of cloud droplets; Entropy (E_n)—quantifying conceptual uncertainty; and Hyper-entropy (H_e)—measuring entropy uncertainty. Cloud droplets are generated through the (E_x, E_n, H_e) parameters, whose accumulation forms complete cloud distributions, thereby enabling qualitative–quantitative mapping of evaluation metrics. Cloud similarity is quantified through similarity measures, where closer proximity between state evaluation clouds indicates higher equivalence in classification levels. The evaluation process involves: (1) determining the actual state cloud (E_x, E_n, H_e) and t benchmark clouds (E_{xt}, E_{nt}, H_{et}); (2) using forward cloud generators for qualitative–quantitative mapping; and (3) calculating similarity measures between actual and benchmark clouds to derive final emergency plan evaluation grades.

2.2.2 Computational Methodology of Cloud Model

Two cloud generators, forward and reverse transformations, achieve the bidirectional conversion between qualitative concepts and quantitative metrics.

Forward Cloud Generator

The certainty degree of individual cloud droplets is computed sequentially, with collective certainty measures constituting a stochastic variable.

Step 1: Generate a normally distributed random number y with E_n as mean and H_e as standard deviation.

Step 2: Produce normally distributed random number x using E_x as mean and y as standard deviation.

Step 3: Calculate the Gaussian certainty degree as follows:

$$A = e^{-(x-Ex)^2/(2)^2}. \tag{6}$$

Step 4: Designate x as a cloud droplet with certainty degree A.

Step 5: Repeat iterations until generating K droplets, where minimum requirements are satisfied: $K \geq 9S^2/\Delta^2$, with S^2 denoting sample variance and Δ being the permissible error threshold.

In urban rail transit emergency plan evaluation, the forward cloud generator methodology is implemented to synthesize the resultant evaluation cloud by processing empirical assessment data through systematic computation of their numerical characteristics (E_x, E_n, H_e).

Reverse Cloud Generator

Step 1: Compute the expectation E_x and the $S2$ variance of cloud droplet samples.

Step 2: Calculate the entropy E_n of cloud droplet samples as follows:

$$E_n = \sqrt{\frac{\pi}{2}} \times \frac{1}{N} \sum_{i=1}^{N} |x_i - E_x|. \tag{7}$$

where in Eq. (7):

- N denotes the sample size
- x_i represents the i sample value

Step 3: Compute the hyper-entropy H_e of cloud droplet samples:

$$H_e = \sqrt{|S^2 - E_n^2|} \tag{8}$$

Step 4: The resultant digital characteristics are formalized as the parameter tuple (E_x, E_n, H_e).

In practical implementation for urban rail emergency plan evaluation, the reverse cloud generator methodology synthesizes expert ratings with predetermined indicator weights through systematic computation, thereby enabling the derivation of comprehensive digital characteristics for assessment clouds.

2.3 Emergency Plan Evaluation Methodology

Based on the cloud model theory, each expert's assessment is conceptualized as a cloud droplet within the evaluation indicator system's universe of discourse. The collective assessments from all experts form a comprehensive cloud whose characteristics reflect the emergency plan's evaluation status. The evaluation procedure comprises the following steps [10]:

Step 1: Factor set determination.

The factor set is established based on the predefined urban rail transit emergency plan evaluation indicator system.

Step 2: Evaluation set and cloud configuration.

The evaluation set V, comprising k qualitative assessments, is formulated as:

$$V = \{V_1, V_2, \ldots, V_i\} \tag{9}$$

In Eq. (9), V_k denotes the k qualitative assessment level. Given the lower bound $d_{\min}$ and upper bound $d_{\max}$ of qualitative assessments, the numerical characteristics for each evaluation level cloud are computed as follows:

$$\begin{cases} E_x = \frac{d_{\max}+d_{\min}}{2} \\ E_n = \frac{d_{\max}-d_{\min}}{6} \\ H_e = g \end{cases} \tag{10}$$

In Eq. (10), parameter g controls fuzziness intensity, with $g = 0.5$ adopted in this study.

Step 3: Weight set determination.

The CRITIC method processes expert rating data to derive an indicator weight set W, formulated as:

Step 4: Comprehensive cloud computation.

Reverse cloud generator operations on expert ratings yield the cloud characteristic matrix Y:

$$Y = \begin{bmatrix} E_{x1} & E_{n1} & H_{e1} \\ E_{x2} & E_{n2} & H_{e2} \\ \vdots & \vdots & \vdots \\ E_{xm} & E_{rm} & H_{em} \end{bmatrix} \tag{11}$$

The resultant comprehensive evaluation cloud for urban rail transit emergency plans is determined as:

$$C = W \times Y. \tag{12}$$

Step 5: Result determination.

The forward cloud generator produces evaluation cloud C (E_x, E_n, H_e), which is subsequently compared with preestablished benchmark clouds through similarity analysis.

3 Case Analysis

3.1 Development of Evaluation Indicator System

This study constructs a dual-dimensional indicator system for emergency plan evaluation based on the Federal Emergency Management Agency (FEMA 2020) National Preparedness Goal System and the operational resilience framework for public organizations in Table 1 [13]. Within the comprehensiveness dimension, four observable variables (F11 plan system rationality, F12 risk assessment completeness, F13 feasibility of safeguard measures, and F14 plan system coordination) systematically evaluate the completeness of risk coverage and system architecture in emergency plans. The operability dimension uses observation indicators, including F21 clarity of departmental responsibilities and F22 scientific rationality of procedures, to examine the implementation effectiveness during emergency response. All indicators were developed with reference to mature measurement scales from authoritative literature such as the National Preparedness Goal System and "Operationalizing Resilience" (2020), ensuring the reliability and validity of the evaluation system. This multi-level indicator design preserves theoretical rigor while strengthening empirical foundations, establishing a structured framework for subsequent quantitative assessments.

The formulation of urban rail transit emergency plans serves to establish systematic protocols for: prevention, monitoring, and early warning; emergency preparedness and response operations; post-incident recovery and reconstruction; information reporting and media management. These protocols aim to ensure timely, orderly, and effective mitigation of operational emergencies, thereby minimizing casualties, economic losses, and social disruptions while maintaining transportation safety and public order. A case study was conducted using a metro corporation's emergency plan, with the evaluation framework structured around three core dimensions: comprehensiveness, operability, and cost-effectiveness. The resultant indicator system is presented in Fig. 1.

Through field investigation and analytical integration, the urban rail transit emergency plans were classified into five distinct tiers. The hierarchical classification

Table 1 Indicator source

Latent variable	Target code	Observational variable	Indicator source
Comprehensiveness	F11	The rationality of the plan system	FEMA (2020). "*National preparedness goal and system*"
	F12	Integrity of risk assessment	
	F13	Feasibility of safeguard measures	
	F14	Coherence of emergency plansF14	
Operability	F21	Clarity of departmental responsibilities	"*Operationalizing resilience: a multi-level framework for public service organizations*"
	F22	The science of the response	
	F23	The rapidity of on-site disposal	
Cost-effectiveness	F31	Consumption of emergency supplies	FEMA (2020). "*Benefit–cost analysis (BCA) reference guide*"
	F32	Expenses for contingencies	

framework with corresponding scoring criteria is systematically presented in Table 2.

3.2 Cloud Model Evaluation

Twenty domain experts from metro operators (50%) and academic/research institutions (50%) were recruited to assess secondary indicators, yielding an initial evaluation matrix. Comparative intensity of each indicator was quantified, followed by computation of conflict degree (Eq. 2), comprehensive information content (Eq. 3), and CRITIC-derived weights (Eq. 4), as detailed in Table 3.

Through the implementation of the reverse cloud generator algorithm, the cloud characteristic matrix Y was derived for indicator sets. By incorporating intra-dimensional weights, the dimensional evaluation cloud parameters for the metro corporation's emergency plan were obtained (Table 4). Subsequent integration with dimensional weights yielded the comprehensive evaluation of cloud parameters:

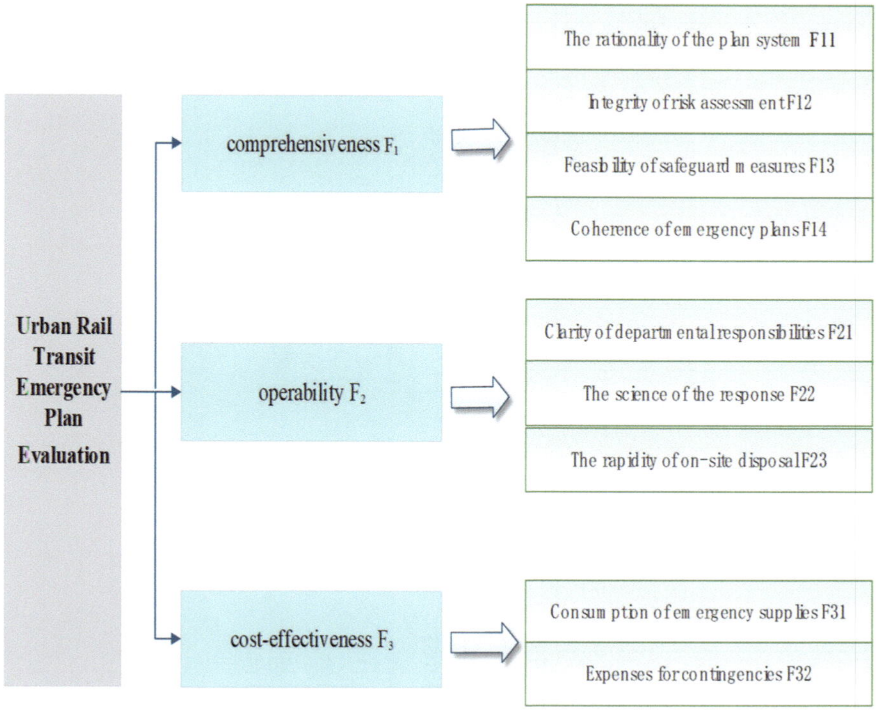

Fig. 1 Urban rail transit emergency plan evaluation

Table 2 Evaluation rating and score

Grade	Excellent	Good	Average	Fair	Poor
Range	[100, 90]	[90, 80]	[80, 60]	[60, 30]	[30, 0]

$$C(E_x, E_n, H_e) = (81.1911, 3.9338, 1.2907)$$

Benchmark cloud parameters derived from Table 1 and Eq. (10) are defined as:

Excellent: (95.0000, 1.6667, 0.5000), Good: (85.0000, 1.6667, 0.5000), Average: (70.0000, 3.3333, 0.5000), Fair: (45.0000, 5.0000, 0.5000), Poor: (15.0000, 5.0000, 0.5000).

3.3 Analysis of Final Evaluation Results

Figures 2 and 3 indicate that the evaluations for the "Comprehensiveness" and "Operability" dimensions fall between the "Good" and "Excellent" levels, demonstrating

Table 3 The weight of the evaluation index

Indicator	Contrast intensity	Conflict degree	Comprehensive information content	Indicator weight	Dimension	Dimension weight	Intra-dimensional indicator weight
F11	1.9468	8.7861	17.1048	0.0702	F1	0.4010	0.1750
F12	3.7320	7.890	29.4455	0.1208			0.3013
F13	3.5253	7.1576	25.2327	0.1035			0.2581
F14	3.5903	7.2303	25.9589	0.1065			0.2656
F21	2.8156	8.3201	23.4261	0.0961	F2	0.2976	0.3230
F22	3.2711	8.8420	28.9231	0.1187			0.3987
F23	2.7821	7.2556	20.1858	0.0828			0.2783
F31	4.3460	8.7961	38.2279	0.1569	F3	0.3013	0.5206
F32	4.8218	7.3006	35.2020	0.1444			0.4794

Table 4 Evaluation results cloud parameters

Dimension	Indicator	Indicator cloud parameters			Dimensional evaluation cloud parameters		
		Exj	Enj	Hej	Exd	End	Hed
F1	F11	94.1000	1.9802	0.2610	88.4007	3.4760	0.7420
	F12	86.1500	3.9667	1.0366			
	F12	86.6500	3.7411	0.9563			
	F14	88.9000	3.6471	0.5165			
F2	F21	86.3500	3.0330	0.9244	85.2042	3.2171	0.3531
	F22	84.0000	3.6346	1.3954			
	F23	85.6000	2.8325	0.3527			
F3	F31	67.7500	4.9506	2.1510	67.6301	5.2510	2.3544
	F32	67.5000	5.5772	2.5753			

satisfactory performance in these two dimensions. Figure 4 shows that the evaluation for the "Cost-Effectiveness" dimension lies between "Fire" and "Average," indicating subpar performance in this dimension. This is primarily due to the lack of specific scoring criteria, insufficient data, and significant uncertainties. Figure 5 reveals that the comprehensive evaluation cloud for the metro company's emergency plan is positioned between "Average" and "Good," suggesting overall deficiencies and substantial room for improvement. Overall, despite the "Average" evaluation result, this emergency plan can still serve as a guideline for incident management, enabling efficient, timely, and orderly responses to emergencies.

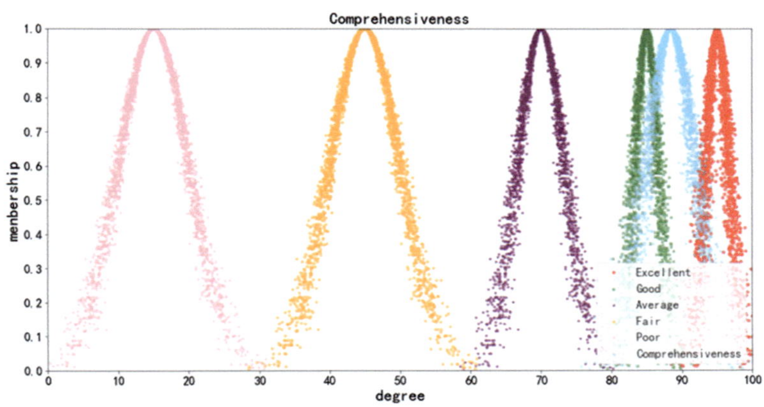

Fig. 2 Emergency plan index cloud and comprehensive evaluation cloud of a subway company

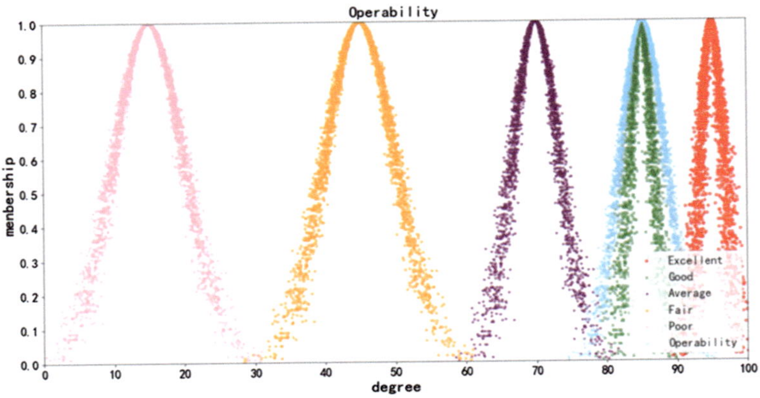

Fig. 3 Emergency plan index cloud and operability evaluation cloud of a subway company

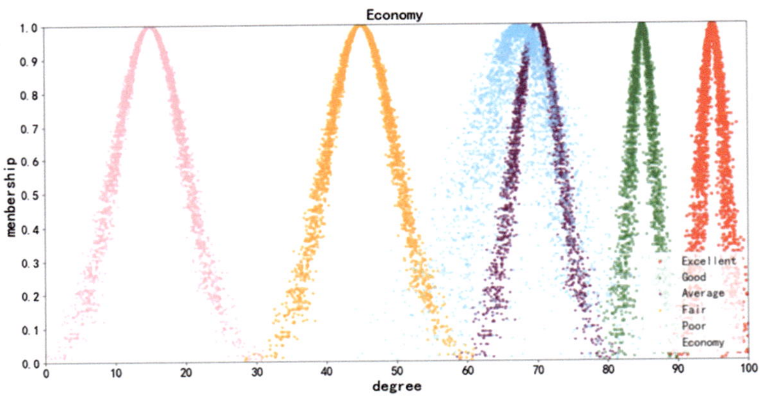

Fig. 4 Emergency plan index cloud and economic evaluation cloud of a subway company

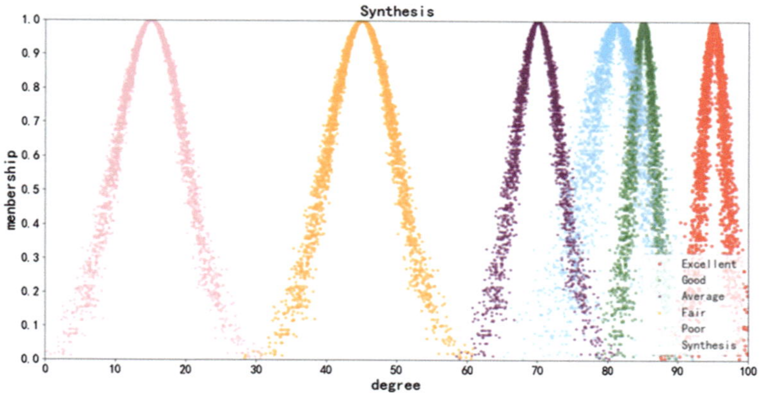

Fig. 5 Emergency plan index cloud and Integrated evaluation cloud of a subway company

4 Conclusion

Scientific research requires reproducibility. Non-reproducible transient phenomena are generally excluded from scientific inquiry. Scientific investigation demands precision, as phenomena lacking quantitative representation often resist mathematical analysis. The proposed cloud model provides a mathematical framework to address these challenges.

This methodology relaxes the restrictive preconditions typically required for studying phenomena conforming to specific probability distributions—conditions that are often difficult to satisfy or even explicitly define. It also resolves the artificial determination of membership degrees in fuzzy set theory by establishing probability density distributions of membership degrees independent of specific conceptual interpretations.

The universality of normal distributions and normal membership functions jointly establishes the theoretical foundation for the pan-applicability of normal cloud models. The pan-normal distribution of cloud X imposes less restrictive assumptions than standard normal distributions (which become its special case), thereby demonstrating broader applicability. Cloud Y introduces stochasticity in membership degrees, where normal membership functions serve as expectation curves independent of concept-specific physical interpretations, better reflecting both the flexibility and regularity of human cognition.

However, to maintain the convenience of the framework and conceptual model, we recommend not adding too many elements to them but finding more specific data analysis methods and models for more accurate quantitative analysis. Otherwise, to ensure the accuracy of the results of the comprehensive method and the applicability of the final proposal to the actual operation, our framework and conceptual model are mainly used for experts and managers with certain urban rail operation experience or related knowledge reserves.

The analysis using the established cloud model validated the feasibility and generalizability of the evaluation method, providing a reference framework for future emergency plan assessments. Compared to methods such as the Analytic Hierarchy Process (AHP) and entropy weighting, the CRITIC method objectively integrates both the contrast intensity and conflict of indicators for weighting, effectively avoiding subjective biases. This makes CRITIC more rational for assigning weights to emergency plan indicators.

References

1. Ju, Y., Wang, A., Liu, X.: Evaluating emergency response capacity by fuzzy AHP and 2-tuple fuzzy linguistic approach. Exp Syst Appl **39**(8), 6972–6981 (2012). https://doi.org/10.1016/j.eswa.2012.01.061
2. Zhang, J., Huang, D., You, Q., Kang, J., Shi, M., Lang, X.: Evaluation of emergency evacuation capacity of urban metro stations based on combined weights and TOPSIS-GRA method in intuitive fuzzy environment. Int J Disast Risk Reduct **95**, 103864 (2023)
3. Luu, C., von Meding, J., Mojtahedi, M.: Analyzing Vietnam's national disaster loss database for flood risk assessment using multiple linear regression-TOPSIS. Int J Disast Risk Reduct **40**, 101153 (2019). https://doi.org/10.1016/j.ijdrr.2019.101153
4. Li, D.Y., Liu, C.Y.: Study on the universality of the normal cloud model. Eng. Sci. **54**(8), 34–35 (2004)
5. Harrod, S.: Capacity factors of a mixed speed railway network. Transp Res E Logist Transp Rev **45**, 830–841 (2009)
6. Orencio, P.M., Fujii, M.: A localized disaster-resilience index to assess coastal communities based on an analytic hierarchy process (AHP). Int J Disast Risk Reduct **3**, 62–75 (2013)
7. Chen, W., Wang, X., Wang, W., Zhu, Y., Cai, Z., Yang, S.: A heterogeneous GRA-CBR-based multi-attribute emergency decision-making model considering weight optimization with dual information correlation. Exp Syst Appl **182**, 115208 (2021)
8. Diakoulaki, D., Mavrotas, G., Papayannakis, L.: Determining objective weights in multiple criteria problems: the CRITIC method. Comput Ops Res **22**, 34–35 (1995)
9. Joosep, H., De Silva, V., Kondoz, A.: Deep multi-critic network for accelerating policy learning in multi-agent environments. Neural Netw. **7**(5), 13–31 (2020)
10. Wu, H.-W., Zhen, J., Zhang, J.: Urban rail transit operation safety evaluation based on an improved CRITIC method and cloud model. J Rail Transp Plan Manag **16**, 100206 (2020)
11. Li, M.-W., Hong, W.-C., Kang, H.-G.: Urban traffic flow forecasting using Gauss–SVR with cat mapping, cloud model and PSO hybrid algorithm. Neurocomputing **99**, 230–240 (2013)
12. Qi, Q., Meng, Y., Zhao, X., Liu, J.: Resilience assessment of an urban metro complex network: a case study of the Zhengzhou metro. Sustainability **14**(18), 11555 (2022). https://doi.org/10.3390/su141811555
13. Federal Emergency Management Agency, FEMA.gov., FEMA.gov. (2020)

Highway Traffic Flow Congestion Warning Model Based on the Weighted Linear Extrapolation Method

Chun Bao, Guoyan Che, Jianqiu Chen, Yanzhi Pang, Benxiao Lou, Mingzhe Chen, Fanglan Tan, Weijie Chen, and Guobin Gu

Abstract Aiming at the problem that the highway management personnel rely on personal experience to open the emergency lane, and it is difficult to scientifically regulate the traffic flow in real time, a highway traffic flow congestion model is proposed. This model processes the video data based on the YOLO-V5s algorithm, extracts the traffic flow parameters, uses polynomial regression to fit their change rules, establishes the model with the common time period index as the training set, and adopts the weighted linear extrapolation method to issue congestion warnings. The results show that for the road sections where the predicted congestion exceeds 30 min, the warning can be issued 10 min in advance. The prediction accuracy rate of the traffic flow density reaches 85.76%, and the prediction accuracy rate of the vehicle speed reaches 88.19%. This provides a new idea for highway congestion warning.

Keywords Traffic engineering · Congestion warning · Weighted linear extrapolation · Traffic flow · Highway

1 Introduction

The number of motor vehicles owned in China continues to increase, but the construction of expressways is difficult to keep pace. Traffic congestion on some sections has intensified, which not only affects the smooth passage but also increases the risk of accidents. Currently, there are many problems in the management and control of

---

C. Bao · J. Chen · Y. Pang · M. Chen · F. Tan · W. Chen · G. Gu (✉)
Guangxi Key Laboratory of International Join for China-ASEAN Comprehensive Transportation, Nanning University, Nanning, China
e-mail: 15251707205@163.com

G. Che
Guangxi Key Laboratory of Intelligent Transportation System (ITS), Guilin University of Electronic Technology, Guilin, China

B. Lou
School of Traffic and Transportation Engineering, Central South University, Changsha, China

K. Subramaniyam and M. Paprzycki (eds.), *Smart Cities and the Internet of Things*, Internet of Things, https://doi.org/10.1007/978-3-032-08290-9_9

expressway emergency lanes [1]. Managers often operate based on personal experience and lack a scientific basis. The control schemes, laws and regulations, as well as the scope of application, all need to be improved. In the research field of traffic flow congestion warning models, numerous scholars have proposed various methods [2]. However, the research on the scientific management and control of expressway emergency lanes is still insufficient [3–5].

2 Data Sources and Preprocessing

The section of the bridge crossing the Nanjing-Hangzhou Expressway in Nanjing is selected, and a 5-km section of the road is regarded as a directed graph, with four monitoring points set up. The You Only Look Once, Version 5, small (YOLO-V5s) algorithm is applied to process the video data. Through steps such as image frame segmentation, feature information extraction, moving target detection, and real-time warning processing, this algorithm can obtain the basic parameters of the traffic flow. After that, the extracted data is cleaned and its integrity is checked to ensure the data quality. The specific steps are as follows.

Step 1: Image frame segmentation: Decompose the video data into image frames at one-minute intervals.

Step 2: Feature information extraction: Utilize the YOLO-V5s [6] network architecture (such as Backbone and Neck) with polynomial regression to fit their change rules, and with the help of the Convolutional Neural Network (CNN) and the Feature Pyramid Network (FPN), extract information such as traffic volume and traffic flow speed.

Step 3: Moving target detection: Construct Anchor boxes to predict the position and size of moving targets, and use the non-maximum suppression technique to eliminate overlapping detection boxes.

Step 4: Real-time warning processing: Rely on the hardware-aware neural network and the new RepVGG-style architecture to reduce latency, adjust the frame rate, improve the recognition accuracy, and achieve warning of road congestion and the management and control the emergency lanes.

Figures 1 and 2 show the processing effects of YOLO-V5s on the video data. To ensure the quality of the data for subsequent research, this paper conducts data cleaning and spatio-temporal integrity inspection on the extracted traffic flow parameters. It has been verified that the original traffic flow data has no missing values, abnormal values, or outliers, and the quality meets the requirements of the research.

Fig. 1 Vehicle rendering of YOLO recognition in scene one

Fig. 2 Vehicle rendering of YOLO recognition in scenario two

3 Data Feature Analysis

Analyze the variation rules of traffic volume, traffic flow density, and traffic flow speed over time, respectively. The variation trends of these three parameters at the four monitoring points are similar. Among them, the cliff-like decline in traffic flow speed and traffic flow density can intuitively reflect the congestion state of the lane.

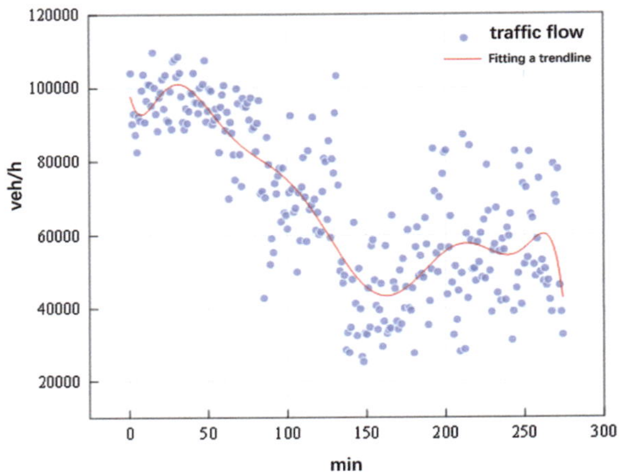

Fig. 3 Fitted plot of traffic volume changes at the first monitoring point

Through comparison, it is found that the traffic flow parameters of each monitoring point show different changes at different time periods. For example, the traffic volume is relatively high in the initial stage and then gradually decreases; the change of traffic flow density is similar to that of traffic volume; and the change of vehicle speed varies among different monitoring points. This indicates that there are complex spatiotemporal evolution rules in traffic flow, and different regions are affected differently by external factors. The details are as follows:

Simultaneous sudden change period of traffic volume: collect parameters such as the number of vehicles and the average speed in minutes, and draw the flow fitting curves of the four monitoring points (as shown in Figs. 3, 4, 5, and 6). By analyzing the slope of the curves, it is found that the slopes of the first three monitoring points change suddenly together in a specific period. Moreover, the traffic volumes of the four monitoring points are all high at the beginning and then decrease. There are regional differences, and the traffic volume of the first monitoring point is the highest.

Simultaneous sudden change period of traffic flow density: draw the density fitting curves of the four monitoring points (as shown in Figs. 7, 8, 9, and 10), and their changes are similar to those of the traffic volume. The initial densities of each monitoring point are different, and they all show a downward trend over time, but the rates of decline and the fluctuation situations are different, which reflects the differences in traffic characteristics in different regions and the influences of external factors.

Simultaneous sudden change period of traffic flow speed: analyze the speed fitting curves of the four monitoring points (as shown in Figs. 11, 12, 13, and 14), and it is found that there are cliff-like sudden changes in the vehicle speed over time. The variation trends and fluctuation ranges of the vehicle speeds at each monitoring point are different. Polynomial regression can well display the variation trend of the traffic flow speed and reveal the evolution process of the traffic flow. Combining the relationship among the three traffic flow parameters and the congestion distribution diagram

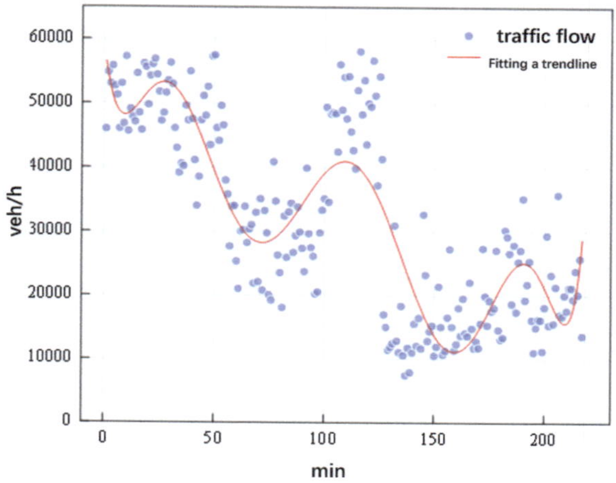

Fig. 4 Fitted plot of changes in traffic flow at the second monitoring point

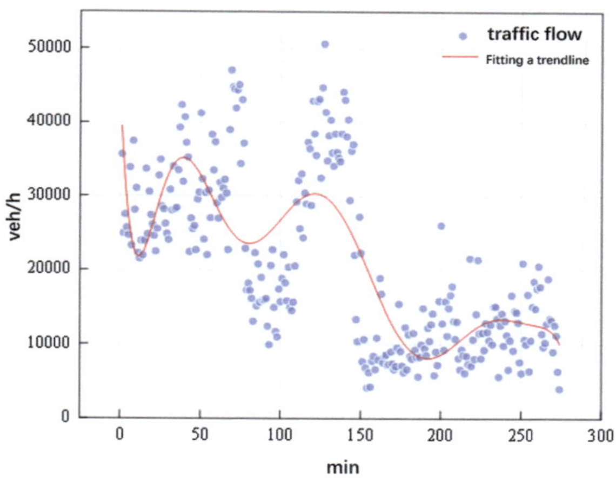

Fig. 5 Fitting of traffic volume changes at monitoring point 3

(Fig. 15), it can be seen that the variation trends of the three parameters are similar, further proving that the change of the traffic flow conforms to the Greenshields model.

4 Traffic Flow Congestion Modeling

To issue a congestion warning for the section between the third and the fourth monitoring points (when the predicted congestion time exceeds half an hour, a warning will be issued ten minutes in advance), the weighted linear extrapolation method is used, and the prediction is carried out through the traffic volume, density, and speed data of the four monitoring points.

Fig. 6 Fitted plot of changes in traffic volume at monitoring point 4

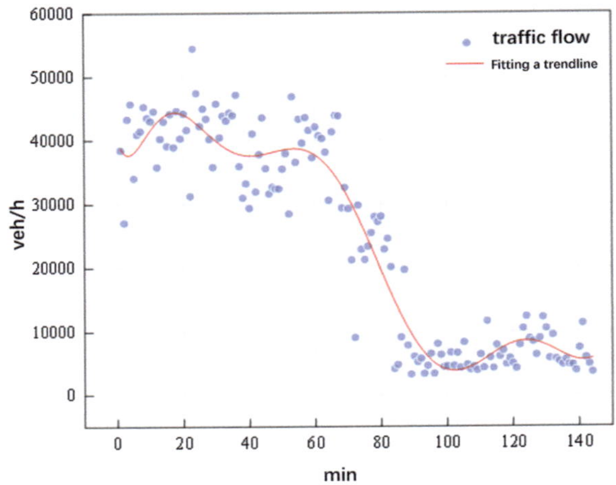

Fig. 7 Fitted plot of changes in traffic density at the first monitoring point

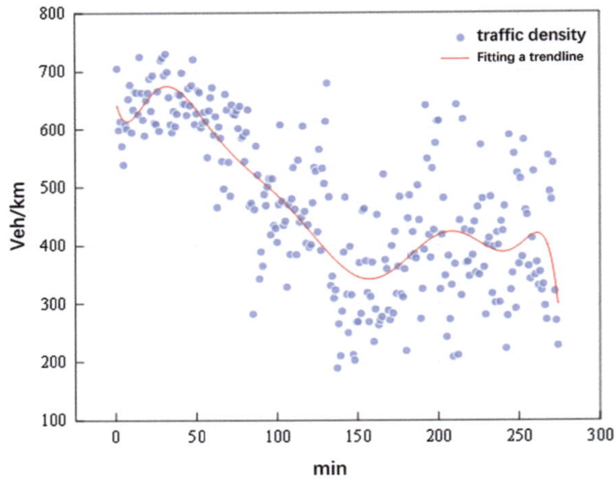

Basic logic of the weighted linear extrapolation model.

Predict the traffic flow parameters (density and speed) of the section between the third and the fourth monitoring points based on the data of the four monitoring points. Predict the future trend and determine whether congestion is likely to occur. If the predicted congestion time exceeds half an hour, the system will issue a warning 10 min before the congestion occurs. The density prediction formula for the section between the third and the fourth monitoring points is shown as formula (1).

$$K_{\text{segment}} = \omega_1 \cdot K_1 + \omega_2 \cdot K_2 + \omega_3 \cdot K_3 + \omega_4 \cdot K_4 \tag{1}$$

where:

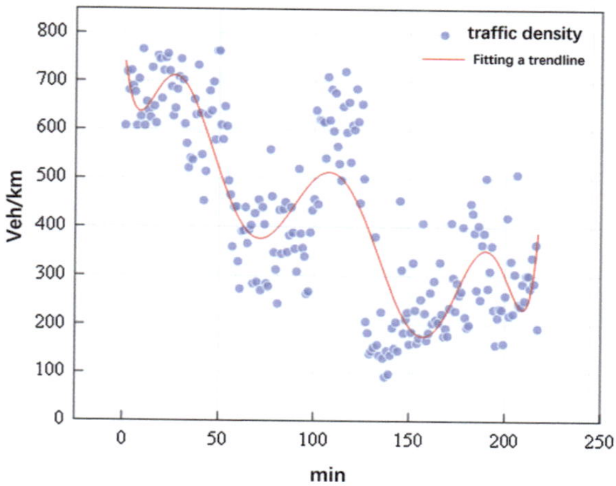

Fig. 8 Fitted plot of changes in traffic density at the second monitoring point

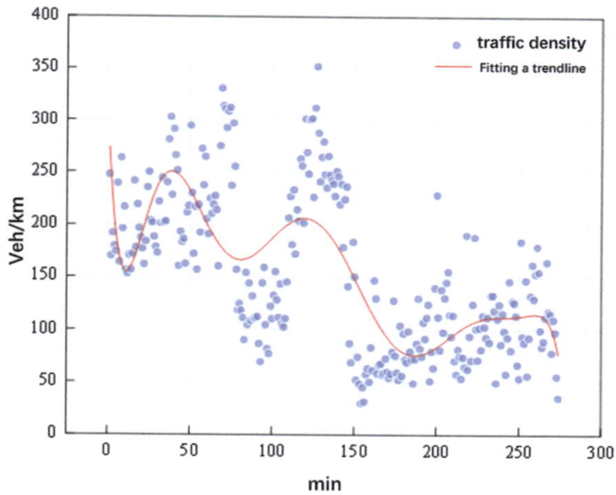

Fig. 9 Fitted plot of changes in traffic density at the third monitoring point

K_{segment} represents the predicted density of the section between the third and the fourth points;

K_{segment} are the densities of the first, second, third, and fourth monitoring points, respectively.

The four weights represent the degrees of influence of the four monitoring points, the number of lanes, and the running length of the traffic flow when making the prediction. According to the distance, the four weights are taken as 0.10, 0.25, 0.30, and 0.35, respectively.

The speed prediction formula for the section between the third and the fourth monitoring points is shown as formula (2).

$$V_{\text{segment}} = \omega_1 \cdot V_1 + \omega_2 \cdot V_2 + \omega_3 \cdot V_3 + \omega_4 \cdot V_4 \tag{2}$$

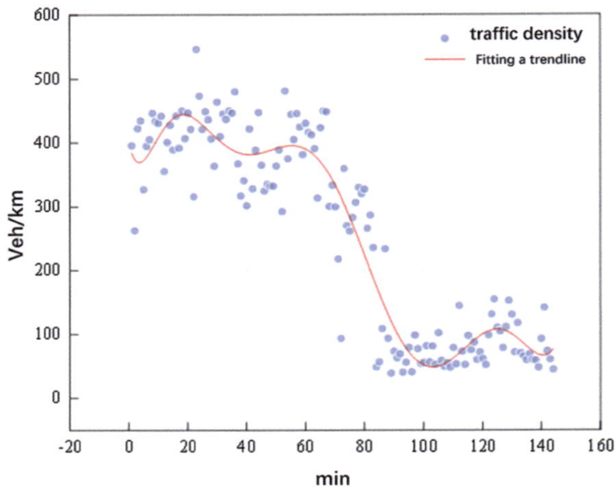

Fig. 10 Fitted plot of changes in traffic density at the fourth monitoring point

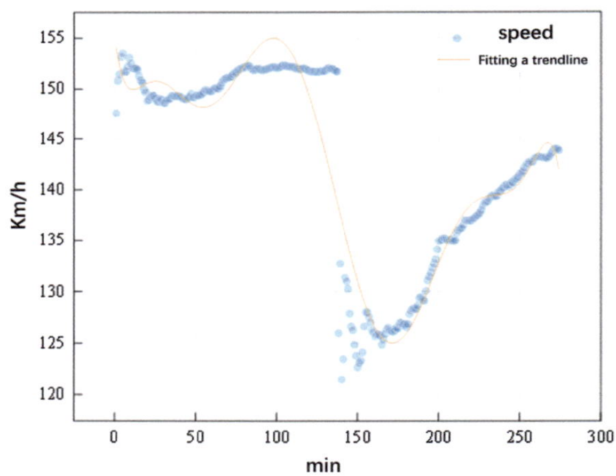

Fig. 11 Fitted plot of traffic speed changes at the first monitoring point

where:

V_{segment} represents the predicted speed of the section between the third and the fourth points;

V_i represent the speeds of the first, second, third, and fourth monitoring points, respectively.

The steps to solve the traffic flow congestion model in this paper are as follows.

Step 1: Set the critical values and parameters of four categories: The critical congestion density D is 0.25 veh/km; the critical congestion speed S is 30.0 km/h; after multiple prediction adjustments, the time threshold T for continuous congestion is taken as 30.0 min; and the time threshold t for advance warning is taken as 10.0 min before the congestion occurs.

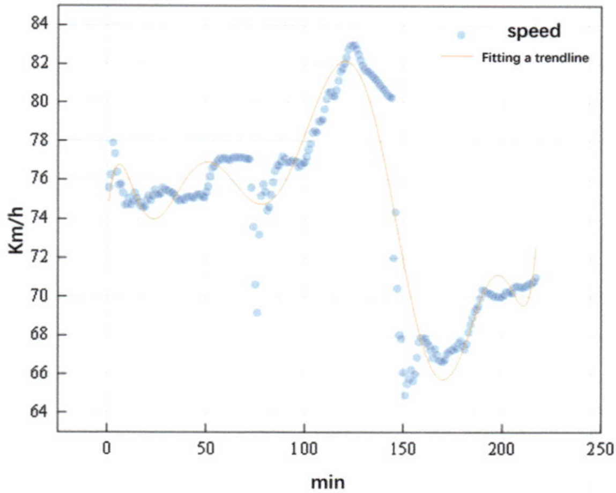

Fig. 12 Fitted plot of changes in traffic speeds at the second monitoring point

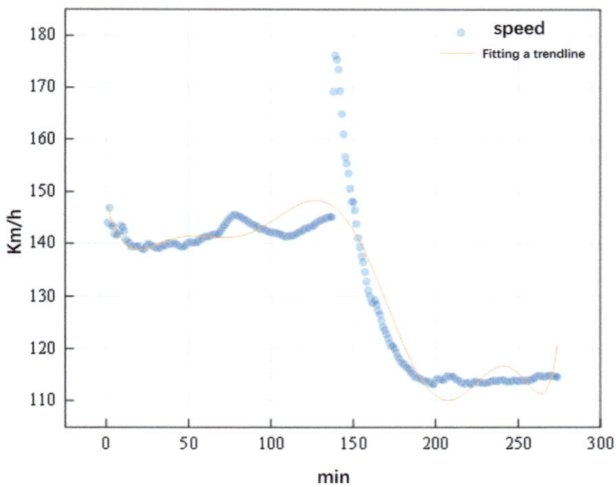

Fig. 13 Fitted plot of vehicle speed changes at the third monitoring site

Step 2: Use the weighted linear extrapolation method to predict the traffic flow density and speed of the section between the third and the fourth points, as shown in formulas (1) and (2), and proportion the weight parameters of the two formulas, and input the corresponding parameters to obtain the weighted density and speed.

Step 3: Based on the weighted traffic flow density and speed, determine the time point of the advance warning through the congestion warning logic.

Step 4: Use the prediction model to check the congestion duration, compare it with the continuous congestion time threshold $T = 30$ min, confirm the time period corresponding to the congestion risk gradient, complete the whole process of the chain reaction of the congestion states of the upstream and downstream lanes, and

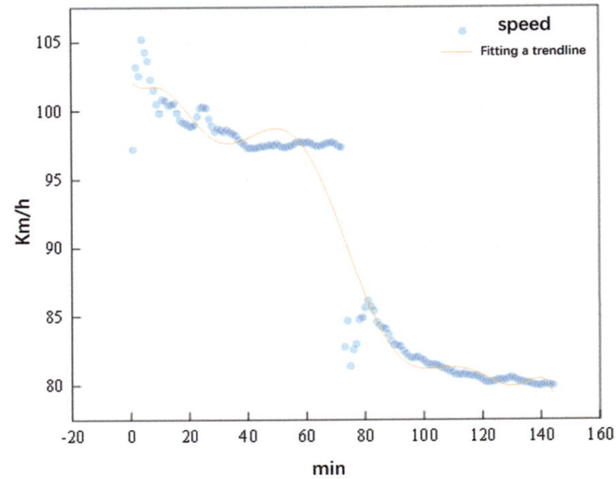

Fig. 14 Fitted plot of vehicle speed changes at monitoring Point 4

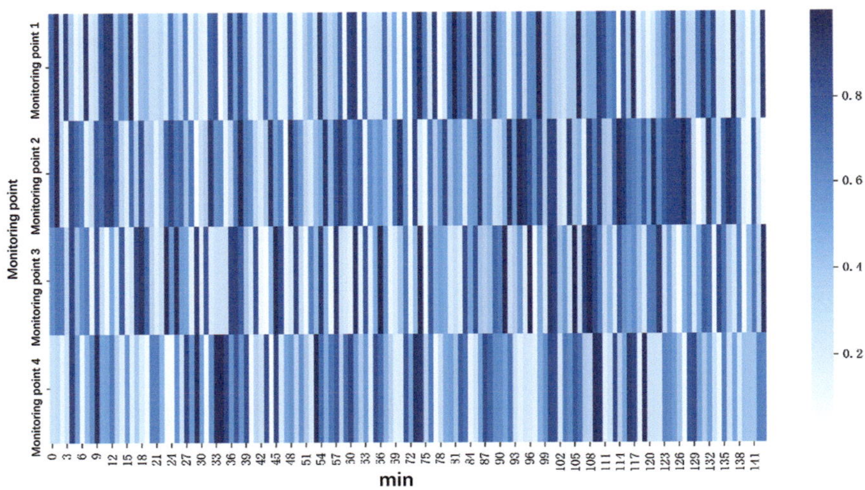

Fig. 15 Distribution of different monitoring points and congestion states

achieve the advance real-time warning of the continuous congestion state of the section between the third and the fourth points.

5 Analysis of Experimental Results

The traffic flow parameter data was re-extracted using YOLO-V5s. Taking the traffic flow density as an example, a scatter plot was drawn for curve fitting to preliminarily judge the matching degree. It was found that there was a significant positive correlation between the predicted and actual traffic flow densities. A scatter plot was drawn to show the corresponding relationship between the predicted and actual densities for consistency evaluation and error analysis. A time trend graph was used to assist in identifying congestion periods. By comparing the trends of the predicted and actual densities, they were found to be consistent, proving the good performance of the model. Finally, the prediction accuracy rate of traffic flow density was 85.76% and that of vehicle speed was 88.19% [7].

6 Conclusion

The model proposed in this study uses video data and polynomial regression to establish a congestion model, and combines the weighted linear extrapolation method to achieve real-time warning, with a high accuracy rate. It provides a new method for expressway congestion warning. In subsequent research, more driving data can be obtained, model parameters can be optimized to improve the prediction accuracy, and the applicability of the model in different traffic environments can be explored.

Funding This research was funded by Guangxi Science and Technology Base and Talent Project for Guangxi Science and Technology Plan Project: Construction of Guangxi Transportation New Technology Transfer Center Platform (No. Guike AD23026029); Construction of Scientific and Technological Innovation Bases for Science Research and Technology Development Plan Project of Nanning City: Intelligent Operation and Maintenance of Transportation Infrastructure and Health Monitoring Platform (No. 20223230).

References

1. Hellinga, B., Van Aerde, M.: An overview of a simulation study of the highway 401 freeway traffic management system. Can J Civil Eng **21**(3), 439–454 (1994)
2. Ha PYJ, Chen S, Dong J et al (2020) Leveraging the capabilities of connected and autonomous vehicles and multi-agent reinforcement learning to mitigate highway bottleneck congestion
3. Yuan, T., Alasiri, F., Zhang, Y., et al.: Evaluation of integrated variable speed limit and lane change control for highway traffic flow. IFAC PapersOnLine **54**(2), 107–113 (2021)
4. Amini M, Hatwagner MF, Mikulai GC et al (2021) An intelligent traffic congestion detection approach based on fuzzy inference system. In: Proceedings of the 2021 IEEE 15th international symposium on applied computational intelligence and informatics (SACI). IEEE, pp 97–104
5. Abood, Z.A., Taher, H.B., Ghani, R.F.: Detection of road traffic congestion using V2V communication based on IoT. Iraqi J Sci **62**, 335–345 (2021)
6. Cremer, M., Papageorgiou, M.: Parameter identification for a traffic flow model. Automatica **17**(6), 837–843 (1981)

7. Rhodes, J.S., Cutler, A., Moon, K.R.: Geometry-and accuracy-preserving random forest proximities. IEEE Trans. Pattern Anal. Mach. Intell. **45**(9), 10947–10959 (2023)

Study on Highway Asphalt Pavement Performance Decay Model in Xinjiang Based on Holt's Exponential Smoothing Prediction Method

Chunmei Liu, Mingyuan Chang, Guobin Zhang, Yihua Wang, and Xiaomin Dai

Abstract The modelling of the performance of asphalt pavement is of great importance in the scientific development of maintenance strategies. The Pavement Condition Index (PCI) and the Pavement Comprehensive Evaluation Index (PCI) are both common pavement performance evaluation indices. The traditional prediction method is difficult to accurately predict the performance of asphalt pavement in Xinjiang, due to the region's unique climate and traffic conditions. This paper proposes an asphalt pavement performance decay modelling method based on Holt's exponential smoothing prediction method, with the PCI and the PQI as the study objects. The Holt double exponential smoothing model is utilised for the purpose of making predictions. The findings indicate that the Holt model exhibits minimal error in short-term prediction, effectively captures the decay trend, and provides data support for pavement preventive maintenance. The study further validates the efficacy of time series modelling in pavement management.

Keywords Asphalt pavement · Performance decay · Holt model · Deterioration condition index (PCI) · Technical condition index (PQI)

C. Liu
Xinjiang Transportation Investment (Group) Co., Ltd, Urumqi, Xinjiang, China

Xinjiang Transportation Investment (Group) Co. Ltd., Operation Branch, Urumqi, Xinjiang, China

M. Chang · G. Zhang · X. Dai (✉)
School of Transportation Engineering, Xinjiang University, Urumqi, China
e-mail: xmdai@xju.edu.cn

Y. Wang
Xinjiang University School of Electrical Engineering, Urumqi, China

M. Chang · G. Zhang · X. Dai
Key Laboratory of Green Construction of Transportation Infrastructure, Intelligent Traffic Control in Xinjiang, Urumqi, China

K. Subramaniyam and M. Paprzycki (eds.), *Smart Cities and the Internet of Things*, Internet of Things, https://doi.org/10.1007/978-3-032-08290-9_10

1 Introduction

Xinjiang is located in the northwestern arid zone, where complex climatic conditions prevail. In addition, the region experiences a high proportion of heavy traffic, which significantly impacts the performance of asphalt pavements. Consequently, the decay law of asphalt pavement in Xinjiang differs considerably from that observed in the central and eastern regions. The extant literature on the subject is predominantly focused on humid or temperate climates, and there is a paucity of decay models for the specific environment of Xinjiang. Conventional prediction methodologies, such as regression models [1] and Markov chains [2], are predicated on the absence of temporal consideration. However, in the context of asphalt pavement deterioration in Xinjiang, temporal variability is a salient factor. Consequently, the efficacy of these models is deemed inadequate for achieving the desired level of prediction accuracy. The grey model [3] applies solely to the prediction of the data trend, which conforms to the exponential law. However, in the context of the complex asphalt pavement condition in the Xinjiang region, the data trend does not conform to the exponential law of the conditions, resulting in poor prediction accuracy. The Autoregressive Integrated Moving Average (ARIMA) model [4] is data-dependent, and this study utilises data from the fifth year, with data from the first 4 years being used for prediction. Given the limited data available, the ARIMA model may not achieve optimal prediction accuracy. The Long Short-Term Memory (LSTM) model [5] has been shown to demonstrate substandard prediction accuracy in circumstances where the data are purely linear trends. The Holt bi-exponential smoothing model has been demonstrated to be a suitable solution for time series data with trend, by dynamically correcting the level and trend components. The present paper adopts the Holt model as an example of the Xinjiang G217 highway, with a view to validating its applicability in the prediction of pavement performance.

2 Asphalt Pavement Performance Research

To predict the trend of asphalt pavement decay with maximum accuracy, many scholars have studied this problem. For example, Mingdy [6] used a grey model to predict the decay of asphalt pavement in the East Xinjiang region, while Yu et al. [7] used an exponential decay model to predict the performance of the Beijing-Chengde Expressway. Liu et al. [8] conducted a study of asphalt pavements in six provinces (namely Guangxi, Gansu, Hunan, Guangdong, Jiangsu and Zhejiang) using the ARIMA model. Peng [9] investigated the application of cumulative series in the context of asphalt pavement decay. Sun Youzhong, meanwhile, studied the prediction of asphalt pavement decay law with a two-parameter model. The climate of the Xinjiang region is characterised by a complex environment, with significant temperature variations between day and night and between winter and summer. These variations result in distinct disease characteristics in the region, leading to accelerated pavement degradation. Additionally, the high proportion of heavy traffic in Xinjiang [10] contributes to the development of ruts and cracks in the pavement [11]. The field of road engineering in China has clearly defined indices for the evaluation of asphalt pavement performance [12]. These include the pavement damage condition index (PCI), the pavement ride quality index (RQI), the pavement rutting depth index (RDI) and the technical condition index (PQI). In the context of Xinjiang, the pavement performance evaluation indices encompass the PCI, RQI, RDI, and PQI. The climate of the Xinjiang region is characterised by a high degree of complexity, with a significant proportion of heavy traffic and considerable temperature variations between day and night. These factors contribute to the increased susceptibility of pavements to cracking and other forms of deterioration [13]. In light of these observations, this study has selected PCI and PQI as the primary predictive indices for investigation.

3 Selection of Samples

The PCI and PQI of the G217 highway were studied by taking sections of the Dushanzi Highway Management Branch, the Karamay Highway Management Branch, the Kuitun Highway Management Branch and the Shaya Highway Management Branch. These sections contain nearly half of the length of the G217 line and differ in climatic factors and traffic volume. The mean values of PCI and PQI were determined for each road section from 2019 to 2023, as illustrated in Tables 1 and 2.

As illustrated in Figs. 1 and 2, there are discernible disparities in the PCI/PQI values of the road sections under the jurisdiction of different administrations, and there are variations in the patterns of decay. Consequently, this study investigates and demonstrates these four groups of road sections separately.

Table 1 Average values for PQI over the last 5 years for the four road segments

Highway management division	2019	2020	2021	2022	2023
Dushanzi	87.49	87.32	87.05	82.64	82.48
Qaramay	88.25	87	85.41	85.1	83.19
Kuytun	89.61	89.22	87.43	86.74	86.47
Shaya	85.54	76.02	73.38	66.59	65.36

Table 2 Average PCI for the last 5 years for four road segments

Highway management division	2019	2020	2021	2022	2023
Dushanzi	87.91	86.56	83.39	82.97	80.94
Qaramay	84.56	82.76	78.74	77.62	75.08
Kuytun	86.25	86.03	83.59	81.66	80.17
Shaya	81.38	71.07	62.92	58.13	53.33

Fig. 1 PCI trends by segment

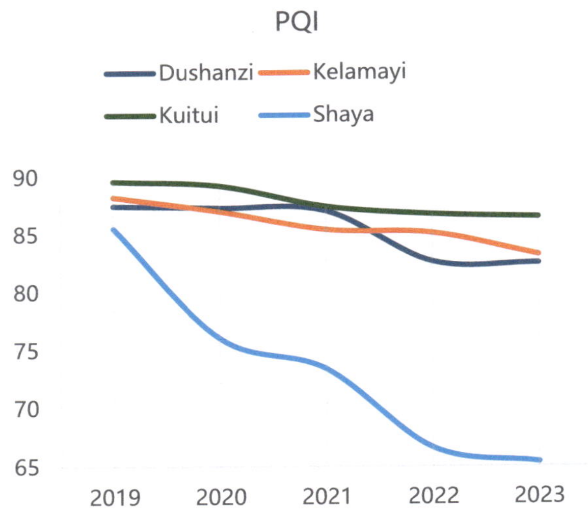

Fig. 2 PQI trends by road segment

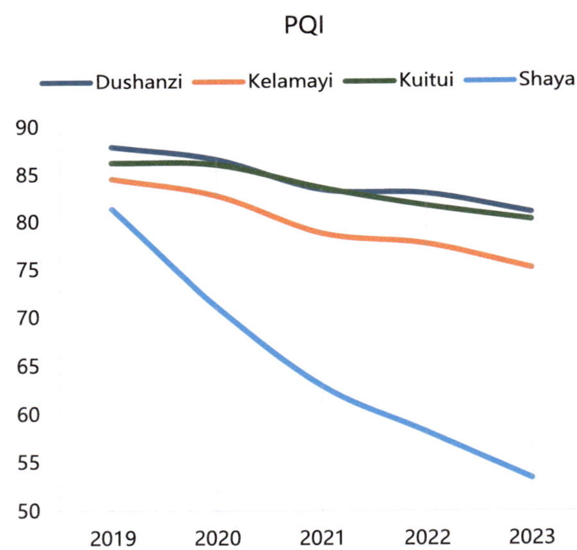

4 Data Preprocessing

The road sections encompassed by the Dushanzi Branch, Klamayi Branch, Kuitun Branch and Shaya Branch were meticulously categorised into eight groups of equal proportion for the study and were assigned numerical designations from A to H. The mean filling method was used to eliminate the mutation values from the data set. The mutation values were substituted with the mean of the data from the year prior to

Table 3 Measured PCI and PQI values by roadway segment

Vintages	2019 data			2020 data			2021 data
Highway management division	Segmentation	PQI	PCI	Segmentation	PQI	PCI	Segmentation
Dushanzi	A	84.778	82.868	A	84.715	78.341	A
	B	92.799	92.799	B	89.003	86.317	B
	……						
Qaramay	A	83.85	81.386	A	78.498	62.832	A
	B	83.85	81.386	B	84.038	76.216	B
	……						
Kuytun	A	93.89	92.648	A	93.444	92.002	A
	B	93.813	92.252	B	93.383	91.631	B
	……						
Shaya	A	88.635	85.653	A	87.012	78.706	A
	B	88.409	83.195	B	87.265	77.187	B
	……						

and the year following the mutation values. This process was undertaken to complete the preliminary data processing. The results of the data processing are presented in Table 3.

5 Holt Bi-exponential Smoothing Forecasting Model

The Holt bi-exponential exponential smoothing model [14] has been shown to exhibit significant advantages in time series forecasting with linear trends. The model's core mechanism is believed to lie in the dual dynamic adjustment of the level and trend values of the series. The model under consideration constructs a recursive parameter updating mechanism by introducing two parameters to be estimated: the level smoothing parameter (α) and the trend smoothing parameter (β). The level component of the model captures the instantaneous state of the series, and the trend component portrays the evolutionary direction of the series. The dual smoothing process of level estimation and trend correction is realised simultaneously through iterative operations, and ultimately, multi-step extrapolation prediction is performed based on the linear combination of the two. This hierarchical modelling strategy has been demonstrated to be an effective method for extracting the long-term trend characteristics of time series data. In addition, it has been shown to enhance the model's capacity to explain complex time series patterns through the adaptive adjustment of parameters. Empirical analysis has revealed that this strategy exhibits enhanced robustness and prediction accuracy. The calculation formulae for this strategy can be found in Eqs. (1), (2) and (3).

Horizontal equation:

$$l_t = \alpha y_t + (1 - \alpha)(l_{t-1} + b_{t-1}) \tag{1}$$

Trend equation:

$$b_t = \beta(l_t - l_{t-1}) + (1 - \beta)b_{t-1} \tag{2}$$

Predictive equation:

$$\hat{y}_{t+h} = l_t + h \cdot b_t \tag{3}$$

In the aforementioned equation, is denotes the level component, is denotes the observation, h denotes the prediction step and α denotes the level smoothing.

parameter in the range of [0, 1], β denotes the trend smoothing parameter in the range of [0, 1], α and β are the optimal solutions derived from the numerical optimisation algorithm of the statsmodels library in Python, and denote the trend component at time t. The above equation is based on a numerical optimisation algorithm.

In the course of the calculations, the initial level component is determined to be the first year's measured value, i.e. 2019's measured data, and the initial trend component is obtained based on the difference of the first 2 years' measured data. Subsequently, the update is gradually followed from the second period, and the observed values of each period and the calculated value of the horizontal component of the previous period are brought into Eq. (1), while the calculated value of the trend component of the previous period and the difference between the calculated value of the horizontal component of the current period and the calculated value of the horizontal component of the previous period are brought into Eq. (2). Upon completion of the update for each period, the predicted value of the subsequent hth period can be calculated by entering the calculated value of the horizontal component and the calculated value of the trend component of the current period into Eq. (3).

6 Parameter Calibration and Model Testing

The selection of appropriate parameters is a critical factor in determining the accuracy of decay model predictions. The Holt bi-exponential smoothing prediction model utilised in this study incorporates two pivotal parameters: the level smoothing parameter (α) and the trend smoothing parameter (β). The statsmodels library in Python is utilised through the grid search algorithm [15]. Initially, the α and β parameters of each section of the route are set to 0.5, and subsequently, they are incorporated into the Holt model for prediction. In the event of a substantial error, the parameters are subjected to adjustment and re-entered into the Holt model until the value

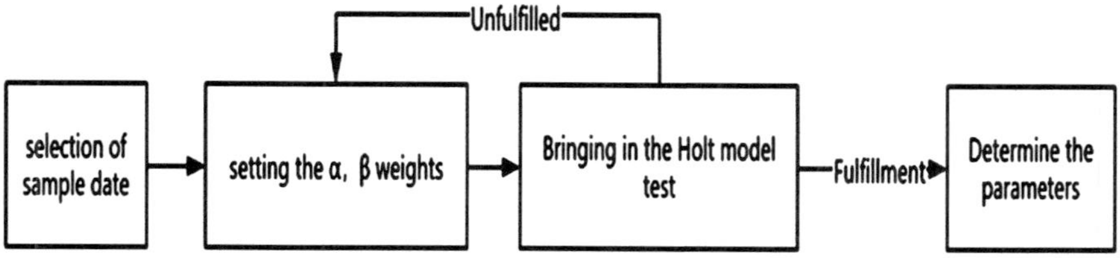

Fig. 3 Calibration flow chart

predicted with the MO parameter is within the error range. The output parameter is then determined, and the process is illustrated in Fig. 3.

The most suitable parameters for each road section are identified individually and subsequently incorporated into the Holt prediction model to calculate the predicted value for each section. The mean absolute error (MAE) and the mean square error (MSE) are then determined, and the results are compiled into a table. A portion of the data is presented in Table 4.

As demonstrated in the above table, it is evident that the α value of each road section is notably elevated, signifying that the recent observations of time series data hold greater significance in the prediction analysis of the current value from the horizontal direction or in sequence prediction. The primary reason for this is that the original series is more clearly indicative of the change over time. Consequently, the current value of the prediction should concentrate more on recent observations. However, it is important to note that there are some sections of the β value that exceed 0.3, which necessitate consideration of the series prediction. The recent data occupy a certain weight, and the trend of the development of the data also occupies a certain ratio. It can be posited that the sequence of change over the entire time period is more gradual, thus resulting in a more accurate prediction analysis. A comparison of the parameters indicates that the prediction of current values is more focused on the observed values in each time period.

In this study, mean absolute error (MAE) and mean square error (MSE) are utilised as performance evaluation metrics for the model. MAE signifies the absolute deviation between the predicted and actual values, with lower values indicating enhanced model prediction accuracy. MSE accentuates the impact of larger deviations through error square operation, and lower values suggest enhanced model prediction stability. In this study, the MAE values obtained from the prediction of PQI are distributed between 0.31 and 0.66, and the MSE values are distributed between 0.02 and 1.96; the MAE values obtained from the prediction of PCI are distributed between 0.45 and 1.75, and the MSE values are distributed between 0.2 and 3.06. The model demonstrates a low prediction error and high accuracy. The sections of Dushanzi Highway Management Sub-bureau and Shaya Highway Management Sub-bureau depicted in the figure correspond to the segments with elevated MAE and MSE values in the data, as illustrated in Figs. 4 and 5. Nevertheless, these values remain within the acceptable error range, thereby validating the model's capacity to execute precise predictions.

Table 4 PCI and PQI predictions, parameters and MAE and MSE values for each road segment

Highway management division	Segmentation	Norm	Actual value	Projected value	α	β	MAE	MSE
Dushanzi	A	PQI	95.523	94.12	0.95	0.25	1.4	1.96
	A	PCI	95.614	94.87	0.92	0.18	0.74	0.55
	B	PQI	88.145	87.23	0.85	0.15	0.92	0.85
	B	PCI	85.708	84.56	0.88	0.12	1.15	1.32
	C	PQI	89.034	88.45	0.89	0.1	0.58	0.34
	C	PCI	86.003	85.12	0.91	0.14	0.88	0.77
	D	PQI	84.606	83.95	0.93	0.2	0.66	0.44
	D	PCI	84.23	83.78	0.87	0.09	0.45	0.2
	E	PQI	84.68	83.21	0.9	0.22	1.47	2.16
	E	PCI	82.215	81.34	0.94	0.17	0.88	0.77
	F	PQI	78.937	77.65	0.96	0.3	1.29	1.66
	F	PCI	75.548	74.92	0.89	0.11	0.63	0.4
	G	PQI	84.272	83.55	0.86	0.13	0.72	0.52
	G	PCI	82.798	81.89	0.92	0.19	0.91	0.83
	H	PQI	84.638	83.47	0.88	0.16	1.17	1.37
	H	PCI	80.882	80.12	0.93	0.21	0.76	0.58
Qaramay	A	PQI	95.329	94.15	0.94	0.27	1.18	1.39
	A	PCI	95.226	93.5	0.91	0.18	1.73	2.99
	B	PQI	82.592	81.34	0.89	0.15	1.25	1.56
	B	PCI	74.318	73.85	0.93	0.22	0.47	0.22
	C	PQI	79.372	78.91	0.95	0.23	0.46	0.21
	C	PCI	70.672	71.23	0.88	0.16	0.56	0.31
	D	PQI	87.355	86.92	0.9	0.19	0.44	0.19
	D	PCI	83.735	82.11	0.96	0.31	1.62	2.62
	E	PQI	88.683	87.45	0.86	0.14	1.23	1.51
	E	PCI	86.411	85.67	0.89	0.12	0.74	0.55
	F	PQI	86.97	85.34	0.92	0.25	1.63	2.66
	F	PCI	83.964	82.55	0.94	0.28	1.41	1.99
	G	PQI	97.07	96.12	0.97	0.33	0.95	0.9
	G	PCI	99.476	98.23	0.98	0.35	1.25	1.56
	H	PQI	91.13	90.45	0.91	0.17	0.68	0.46
	H	PCI	89.026	88.12	0.93	0.2	0.91	0.83
Kuytun	A	PQI	93.007	92.88	0.82	0.08	0.13	0.02
	A	PCI	91.927	91.45	0.85	0.11	0.48	0.23
	B	PQI	92.582	92.34	0.78	0.06	0.24	0.06

(continued)

Table 4 (continued)

Highway management division	Segmentation	Norm	Actual value	Projected value	α	β	MAE	MSE
	B	PCI	90.575	90.21	0.81	0.09	0.36	0.13
	C	PQI	93.804	93.55	0.87	0.12	0.25	0.06
	C	PCI	92.284	91.87	0.89	0.14	0.41	0.17
	D	PQI	93.701	93.45	0.84	0.1	0.25	0.06
	D	PCI	92.228	91.92	0.86	0.13	0.31	0.1
	E	PQI	91.615	90.34	0.93	0.25	1.28	1.64
	E	PCI	89.312	88.12	0.95	0.22	1.19	1.42
	F	PQI	85.497	84.23	0.96	0.31	1.27	1.61
	F	PCI	81.164	80.55	0.88	0.18	0.61	0.37
	G	PQI	94.271	93.12	0.9	0.2	1.15	1.32
	G	PCI	95.799	94.45	0.92	0.24	1.35	1.82
	H	PQI	83.604	82.34	0.94	0.27	1.26	1.59
	H	PCI	76.955	76.23	0.91	0.19	0.73	0.53
Shaya	A	PQI	68.277	67.45	0.96	0.31	0.83	0.69
	A	PCI	58.017	57.23	0.97	0.35	0.79	0.62
	B	PQI	78.783	77.12	0.95	0.29	1.66	2.76
	B	PCI	61.862	60.34	0.98	0.37	1.52	2.31
	C	PQI	55.275	56.91	0.9	0.3	1.63	2.66
	C	PCI	51.967	53.21	0.97	0.35	1.24	1.54
	D	PQI	50.015	51.34	0.99	0.4	1.33	1.77
	D	PCI	45.952	47.21	0.96	0.4	1.26	1.59
	E	PQI	83.929	82.45	0.93	0.25	1.48	2.19
	E	PCI	76.866	75.34	0.94	0.28	1.53	2.34
	F	PQI	81.278	80.12	0.89	0.2	1.16	1.35
	F	PCI	82.985	81.23	0.91	0.23	1.75	3.06
	G	PQI	77.204	76.55	0.87	0.18	0.65	0.42
	G	PCI	77.084	76.12	0.9	0.22	0.96	0.92
	H	PQI	78.51	77.34	0.95	0.3	1.17	1.37
	H	PCI	70.102	69.45	0.92	0.25	0.65	0.42

It is evident from the MAE and MSE of the PCI value of the road section of Karamay Management Highway Branch and the MAE and MSE of the PQI of the road section of Shaya Management Highway Branch that there are some road sections where the MSE value exceeds the MAE value, and some road sections where the MAE and MSE values are extremely close to each other. The phenomenon that the MSE value is substantially larger than the MAE value is also apparent. The absence of

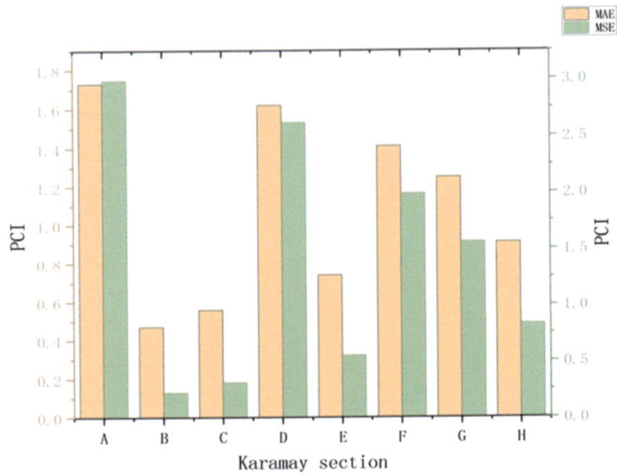

Fig. 4 Histogram of MAE versus MSE for PCI in the Klamath region

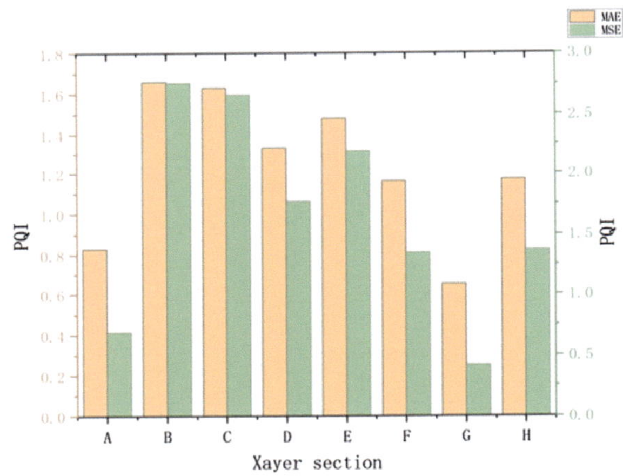

Fig. 5 Histogram of MAE versus MSE for PQI in Shaya region

the MAE value is indicative of the non-existence of exceedingly erroneous samples within the prediction data. This, in turn, signifies that the prediction model in question possesses both a high degree of accuracy and credibility. Furthermore, the model demonstrates a high level of accuracy and credibility, thus resulting in a prediction value that is regarded as high from a comprehensive standpoint.

7 Conclusion

This paper uses the Xinjiang G217 highway as a case study. By collating the measured data of the G217 highway from 2019 to 2023, the Holt bi-exponential smoothing prediction model of PQI and PCI of asphalt pavement was successfully constructed.

The model parameters of different road sections were then determined, and a reliable prediction model was finally obtained. This study is not without its limitations. Firstly, the selection of sections of the Xinjiang G217 highway was somewhat arbitrary, and secondly, the climatic conditions, traffic volume [10] and materials used in asphalt pavements [16] and other elements in other regions are quite different. It is therefore necessary for the applicability of the model to other regions to be ascertained through further testing and verification.

Acknowledgements The present study received financial support from the Xinjiang Transportation Investment Group Supported Project (grant number: EKXFWCG2024040202), the Xinjiang Natural Science Foundation (grant numbers 2024D01A53) and the Xinjiang Uygur Autonomous Region "Dr. Tianchi" Project. The Foundation (grant numbers 2024D01A53) and the Xinjiang Uygur Autonomous Region "Dr. Tianchi" Project have been identified as the relevant bodies.

References

1. Lei, W., Li, J., Ouyang, Z., et al.: A study on risk assessment and prediction of bovine tuberculosis based on multivariate logistic regression model. Chin. J. Animal Infect. Dis. **22**, 1–10 (2025)
2. Li, Y.: Prediction of asphalt pavement overhaul timing based on Markov chain. Highway Motor Transp. **15**(03), 163–166 (2019)
3. Yao, J.-H., Zhao, J., Du, J., et al.: Research on the performance of continuously reinforced concrete composite asphalt pavement based on gray Markov model. Highway Motor Transp. **40**(01), 61–64 (2024)
4. Lu, Z., Yang, C., Li, J., et al.: Vehicle speed condition prediction based on ARIMA-LSTM and RBF-NOA. Spec. Veh. **15**(04), 45–48 (2025)
5. Ren, W., Liu, J., Deng, Y., et al.: Research and prediction of air pollution coupling based on RF regression and LSTM neural network: a case study of Beijing. Sci. Technol. Innov. **15**(10), 22–26 (2025)
6. Mingdy, W.: Analysis of the service performance and decay prediction of asphalt pavement in the east Xinjiang region. Heilongjiang Transp. Sci. Technol. **48**(02), 31–35 (2025)
7. Yu, B.-H., Gao, Y.-M., Yang, W.-Y., et al.: Analysis of asphalt pavement performance decay law of Beijing Jingcheng expressway. Shandong Transp. Sci. Technol. **9**(05), 1–9 (2021)
8. Liu, D., Chen, J., Liu, H.: Prediction of asphalt pavement performance based on ARIMA model. Hunan Transp. Sci. Technol. **50**(02), 18–24 (2024)
9. Peng, Z.: Research on asphalt pavement performance decay model based on cumulative series. Chin. For. Highway **38**(05), 57–61 (2018)
10. Yu, H., Zhang, X., Pu, X., et al.: Study on asphalt pavement deterioration law of common national highway considering the influence of cumulative freight traffic. Tianjin Constr. Sci. Technol. **34**(05), 5–7 (2024)
11. Yan, J., Song, W., Sun, X.: Research on the causes of rutting and anti-rutting technology of asphalt pavement of municipal trunk road. Sci. Technol. Innov. **65**(09), 113–116 (2025)
12. Zhong, W., Yang, J., Yan, Y.: Analysis of structural performance evaluation indexes of highway old asphalt pavement. Highway Transp. Sci. Technol. **16**(02), 138–140 (2020)
13. Zhou, R., Guo, X., Zhao, Y., et al.: Temperature analysis of asphalt mixture construction under large temperature difference and windy environment. Heilongjiang Transp. Sci. Technol. **48**(03), 1–11 (2025)
14. Ma, L., Yang, W., Ma, X., et al.: Application of SARIMA model and Holt-Winters exponential smoothing model in the prediction of HIV/AIDS cases in Xinjiang production and construction corps. J. Prevent. Med. Intell. **40**(11), 1339–1345 (2024)

15. Wen, B., Dong, W., Xie, W., et al.: Random forest parameter optimization based on improved grid search algorithm. Comput. Eng. Appl. **54**(10), 154–157 (2018)
16. Yang, Y.: Research on modification technology of asphalt pavement materials and evaluation of road performance. Sichuan Cement **87**(02), 145–147 (2025)

Research on Electric Vehicle Scheduling Methods Considering the Uncertainty of Users' Scheduling Willingness

Ran Lyu, Mingxing Guo, Li Lan, Yong Zhang, Su Wang, and Zikang Ding

Abstract With the booming development of the electric vehicle industry, the access of a large number of electric vehicles to the power grid will have a profound impact on the power system. The disorderly charging behavior of large-scale electric vehicles will exacerbate the peak–valley load difference of the system, bringing potential risks to the stable operation of the power system. On the contrary, if we can effectively guide the orderly charging of electric vehicles, we can not only reduce the negative impact of the charging load on the power system but also play a positive role in shaving peaks and filling valleys and improve the system operating efficiency. This chapter focuses on the large number of electric private car groups, in-depth analysis of their charging behavior characteristics, and scheduling response psychology; constructs a hierarchical scheduling architecture; and fully exploits the scheduling potential of electric private cars through the incentive compensation mechanism to realize the optimal regulation of load. Based on the actual demand, this chapter effectively alleviates the impact of large-scale electric vehicle access to the power grid on the system, improves the social benefits, and has strong practical significance.

R. Lyu · M. Guo · L. Lan · Y. Zhang · S. Wang
State Grid Shanghai Electric Power Economic and Technology Research Institute, Shanghai, China
e-mail: lvran.deyouxiang@163.com

M. Guo
e-mail: mingxing_guo@163.com

L. Lan
e-mail: ll_lan@126.com

Y. Zhang
e-mail: 13818015695@163.com

S. Wang
e-mail: 776132121@163.com

Z. Ding (✉)
School of Electrical Engineering, Southeast University, Nanjing, China
e-mail: ding_zikang@163.com

K. Subramaniyam and M. Paprzycki (eds.), *Smart Cities and the Internet of Things*, Internet of Things, https://doi.org/10.1007/978-3-032-08290-9_11

Keywords Electric vehicle · Orderly charging · Scheduling willingness · Load dispatch · New energy

1 Introduction

In recent decades, rapid global economic development has triggered serious energy and environmental problems. The massive consumption of non-renewable energy sources, especially coal and oil, has resulted in the emission of large quantities of polluting gases, exacerbating environmental problems. Fuel-fired automobiles have caused tremendous pressure on the global environment and energy, and their consumption of large amounts of petroleum resources while emitting large amounts of exhaust gases has become one of the main sources of air pollution, and it is urgent to solve the pollution problems caused by traditional automobiles [1–6].

Electric vehicles (EVs) are green, efficient, and low consumption, and they excel in solving global environmental problems. Therefore, countries around the world are vigorously developing new energy vehicles and introducing various policies to promote their development. The United States has invested a lot of money in building charging stations and other facilities; many European countries have implemented measures such as high subsidies, low vehicle taxes, and high emission standards; Japan has implemented measures such as lowering the purchase price of cars and providing financial subsidies to charging stations. China also attaches great importance to the development of new energy vehicles, put forward the strategic significance of the development of new energy vehicles in 2020. Under the incentives of the above measures, new energy vehicles have been developing rapidly [7–9].

The disorderly charging behavior of electric vehicles will bring impact to the power grid, and the implementation of scientific and reasonable orderly charging scheduling strategy can significantly reduce the negative impact of electric vehicle charging load on the power grid [8–12].

2 Electric Vehicle Users' Willingness to Scheduling Model

2.1 Electric Vehicle Aggregator Regulation Strategies

The scheduling strategy adopted by the EV aggregator is a layered scheduling model, which subdivided the entire scheduling architecture into three layers, namely the distribution control center, the aggregator, and the charging piles, as illustrated in Fig. 1, where the distribution control center is responsible for releasing the day-ahead prices and real-time prices; the aggregator is responsible for making day-ahead price plans and announcing the prices for electricity sales and incentive compensation, and the charging piles make and execute individual EV charging plans based on the users'

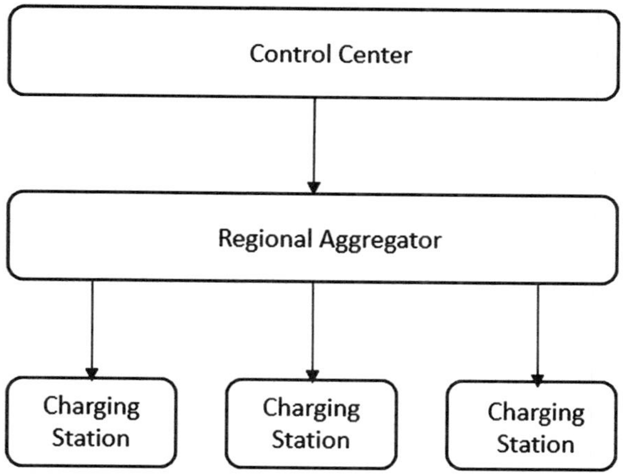

Fig. 1 Electric vehicle aggregator scheduling architecture

scheduling wishes and price information [10–13]. The aggregator is responsible for setting the day-ahead power purchase plan and announcing the power sales price and incentive compensation price, and each charging post sets and executes the charging plan for individual EVs based on the user's scheduling willingness and the power price information. The specific scheduling architecture is shown in Fig. 1.

2.2 *Factors Influencing Electric Vehicle Group Scheduling Intentions*

The willingness of the EV community to scheduling is mainly influenced by the following two factors: scheduling time and the price of incentive compensation [13, 14].

Scheduling time is one of the important factors affecting the willingness of EV users to schedule. Define the proportion of time that users are involved in scheduling as λ:

$$\lambda = \frac{t_d}{t_g} \tag{1}$$

where t_d is the time the EV user participates in the scheduling and t_g is the time the EV is on the power grid.

In this chapter, some questionnaires are released to investigate the effect of λ on the scheduling willingness w_t of electric vehicle users. Through the results of the questionnaire, the following conclusions can be obtained: When the scheduling time t_d is less than half of t_g, the user's willingness to schedule is very high and is not affected by the length of the scheduling time; when the scheduling time t_d exceeds half of t_g, the user's willingness to schedule decreases rapidly with the increase in

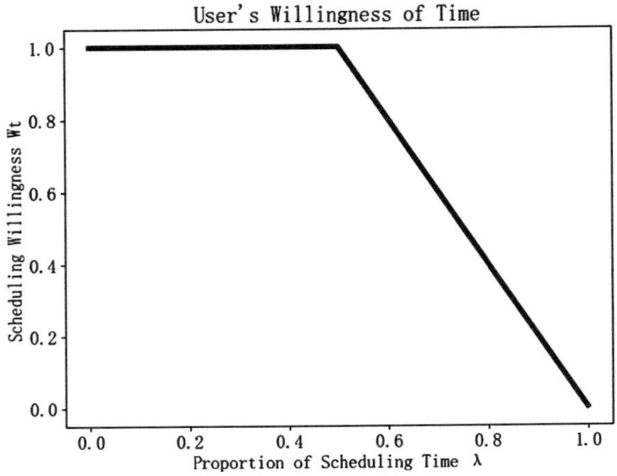

Fig. 2 Electric vehicle user scheduling time willingness curve

the scheduling time. The user scheduling time willingness curve for electric vehicles is obtained in Fig. 2 by fitting.

Incentive compensation price is also one of the important factors affecting users' willingness to schedule; in general, the higher the incentive compensation price, the higher the users' willingness to schedule, and accordingly, the higher the proportion of a user group willing to receive scheduling [15]. In order to quantify the influence of incentive compensation price on the scheduling willingness of user groups, this chapter adopts the Weber–Fechner law to describe the relationship between the two.

The Weber–Fechner law is an important concept in psychology and physiology that reveals the relationship between mental and physical quantities. In this chapter, the Weber–Fechner law is used to model the degree of response of the EV user group to the incentive compensation, when the scheduling incentive compensation issued by the EV aggregator is π yuan/h, the proportion of users choosing load scheduling to the total number of users is x, and defining the charging scheduling user responsiveness function $x = f(\pi)$; then, we have:

$$x = \begin{cases} 0, & 0 < \pi < \pi_0 \\ k\ln(\pi) + b, & \pi_0 \leq \pi < \pi_{\max} \\ x_{\max}, & \pi \geq \pi_{\max} \end{cases} \tag{2}$$

where π_0 and $\pi_{\max}$ are price thresholds; π_0 shows that only when the incentive compensation price reaches a certain value, the user group has the willingness to scheduling; $\pi_{\max}$ shows that when the price reaches a certain level, even if the compensation price is increased, the proportion of the user group's scheduling will not increase.

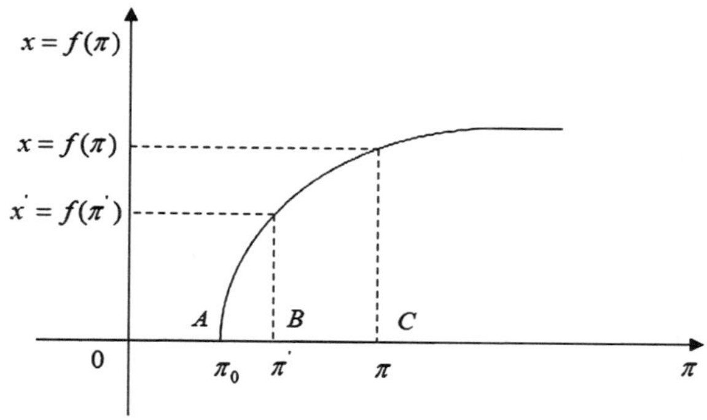

Fig. 3 Plot of updated user responsiveness function

2.3 Electric Vehicle User Utility Model

The price determines the proportion of the population that responds when the user's scheduling time probability function is.

The user group scheduling time willingness affects the user's perception of the scheduling compensation price [16, 17], which defines the user group scheduling utility, defined by the following equation:

$$\pi\prime = w_t \cdot (\pi - \pi_0) + \pi_0 \tag{3}$$

where π_0 is the price threshold, the interpretation of the effect on the willingness to schedule time is shown in Fig. 3.

As shown above, the price threshold π_0 is at point A in the figure, the incentive compensation price π is at point C, the utility $\pi\prime$ is at point B, and the user's willingness to schedule is w_t. Thus, the charge/discharge scheduling responsiveness function of the user group is updated to $x\prime = f(\pi\prime)$.

3 Electric Vehicle Aggregator Charge and Discharge Scheduling Model

3.1 Revenue of Electric Vehicle Aggregator and Electric Vehicle Users

The final revenue of an EV aggregator is as follows: basic power sales revenue + discharge revenue—power purchase costs—costs due to scheduling. The specific revenues and expenses of an EV aggregator, for example, for charge/discharge scheduling, are as follows:

$$\sum_{N=1}^{N_{\max}}\left\{n_{\min}\cdot ec\cdot\eta_c\cdot c_{\text{sell}}\cdot B+\sum_{n=n_a}^{n_a+n_{all}-1}\cdot B-n_{\text{add}}\cdot\pi_d/4\right\} \tag{4}$$

Final revenue for EV users: revenue due to scheduling - basic power purchase cost; the specific revenues and expenditures for EV users are as follows:

$$n_{\text{add}}\cdot\pi_d/4-n_{\min}\cdot ec\cdot\eta_c\cdot c_{\text{sell}}\cdot B \tag{5}$$

3.2 *Development of Charging and Discharging Plans for Electric Vehicles*

The charging and discharging schedules for EV users participating in the charging and discharging scheduling are as follows.

The charging and discharging schedule objective function is as follows:

$$\min\sum_{n=\text{na}}^{\text{na}+\text{nall}-1}\left[cc(n)\cdot ec\cdot c_{\text{ahead}}(n)-cd(n)\cdot ed\cdot c_{\text{real}}(n)\right]\cdot B \tag{6}$$

The constraints of the charging plan are as follows:

(a) The SOC at the end of the scheduling time is >0.95 and ≤1.

$$\left\{\sum_{n=\text{na}}^{\text{na}+\text{nall}-1}\left[cc(n)\cdot ec\cdot\eta_C-cd(n)\cdot ed/\eta_d\right]\right\}>0.95 \tag{7}$$

$$\left\{\sum_{n=\text{na}}^{\text{na}+\text{nall}-1}\left[cc(n)\cdot ec\cdot\eta_C-cd(n)\cdot ed/\eta_d\right]\right\}\leq 1 \tag{8}$$

(b) SOC is >0.2 and <1 before the start of each scheduling period.

$$\left\{\sum_{n=\text{na}}^{\text{ns}}\left[cc(n)\cdot ec\cdot\eta_C-cd(n)\cdot ed/\eta_d\right]\right\}\geq 0.2 \tag{9}$$

$$\left\{\sum_{n=\text{na}}^{\text{ns}}\left[cc(n)\cdot ec\cdot\eta_C-cd(n)\cdot ed/\eta_d\right]\right\}\leq 1.0 \tag{10}$$

(c) *cc*, *cd* cannot be 1 at the same time for each scheduling period.

$$\sum_{n=\mathrm{na}}^{n=\mathrm{na}+\mathrm{nall}-1} \{[cc(n) + cd(n)] \leq 1\} \tag{11}$$

4 Example and Analysis

4.1 Real-Time Power Prices and Base Loads

The real-time power price [18] and the base load of a residential area [19] used in this chapter are shown in Figs. 4 and 5.

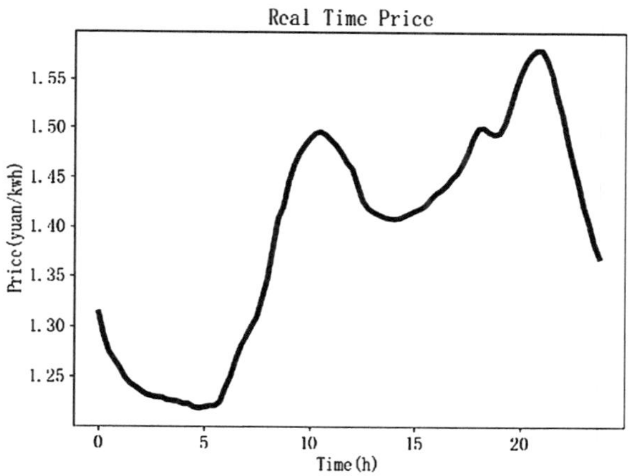

Fig. 4 Real-time price curve

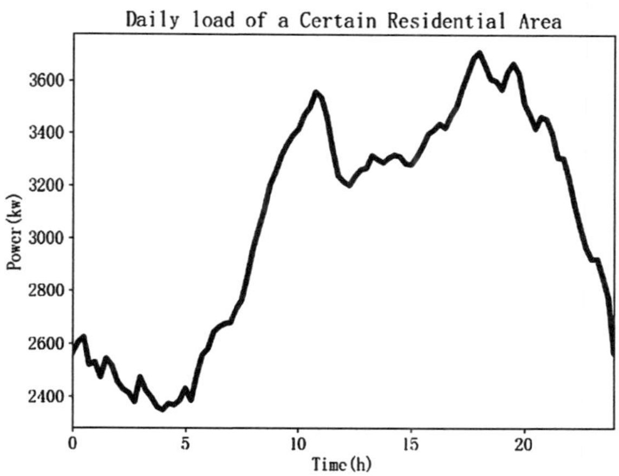

Fig. 5 One day's base load for a certain residential area

Assuming that the day-ahead electricity price is the real-time electricity price minus 0.6 yuan/kWh, and in order to protect the rights and interests of EV users, it is stipulated that the price of electricity sold by the EV aggregator shall not be higher than the maximum value of the day-ahead price, so the price of electricity sold by the EV aggregator is 1.0 yuan/kWh.

4.2 Electric Vehicle Charging and Discharging Load Scheduling

The electricity sales price is 1.0 yuan/kWh, and the charging and discharging compensation price is 0.70 yuan/h for the residential area and 0.84 yuan/h for the work area; 500 private cars in the residential area and 125 private cars in the work area were selected.

Charge/discharge scheduling is used, and the scheduling results are as follows:

(a) Total electric vehicle load curve diagram.
(b) Total load curve diagram.
(c) Load adjustment rate curve diagram.

Specific benefits for electric vehicle aggregators are listed below.

From Fig. 6, it can be seen that the charging and discharging scheduling significantly reduces the peak charging load of electric private cars during the two peak periods of electricity consumption from 8:00 to 10:00 and from 17:00 to 21:00; from Figs. 7 and 8, it can be seen that the load adjustment rate of the charging and discharging scheduling reaches the highest of 4.0% during the peak period of electricity consumption from 18:00 to 21:00, and during the trough period from 22:00 to 4:00 the next day, the load adjustment rate of the charging and discharging scheduling reaches the highest of 3.0%, indicating that the charging and discharging scheduling has a significant effect on filling the valley; as shown in Table 1, the charging and discharging scheduling of private cars in the residential area adds 44.4% to the aggregator's revenue, and the charging and discharging scheduling of private cars in the work area adds 40.2% to the aggregator's revenue.

From the above, it is clear that EV charging and discharging scheduling can bring benefits to users and aggregators while reducing the peak-to-valley difference in system load.

5 Conclusion

This chapter explores the load scheduling potential of electric vehicles by establishing an electric vehicle charging scheduling model. Firstly, it establishes the electric vehicle aggregator scheduling architecture, determines the aggregator scheduling strategy and scheduling means, that is, the scheduling strategy of hierarchical

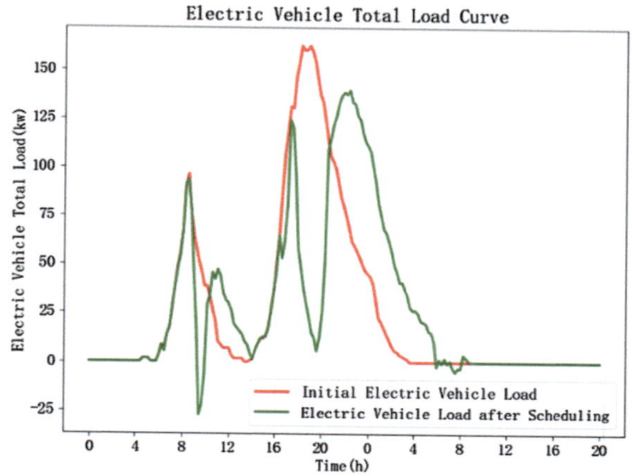

Fig. 6 Total load curve of EVs for scheduling

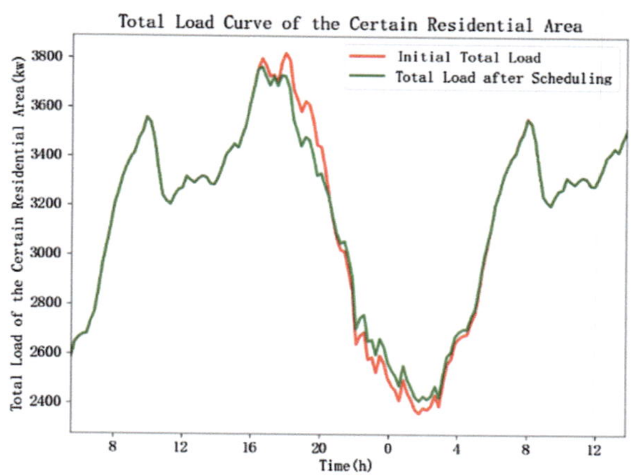

Fig. 7 Total load curve of a certain residential area for scheduling

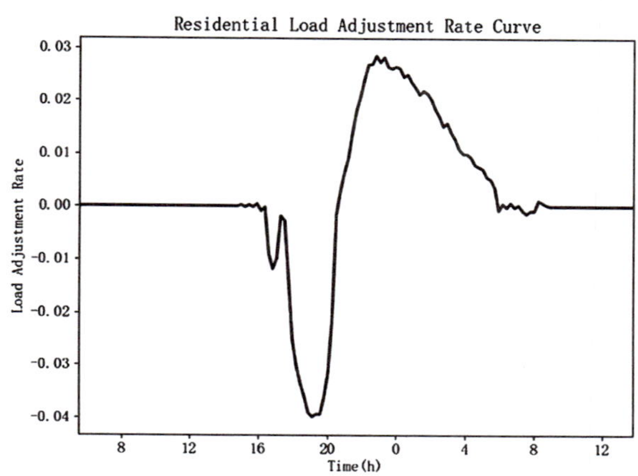

Fig. 8 Scheduling load adjustment rates

Table 1 Electric vehicle aggregator scheduling gains

	Workspace	Residential area
Unordered charging revenue (yuan)	198.89	675.42
Orderly charging revenue (yuan)	278.75	975.35
Scheduling revenue (yuan)	79.86	299.93

scheduling and the scheduling means of incentive compensation, and then establishes the scheduling willingness model of the electric private car users; the scheduling willingness model mainly takes into account the influence of the price of incentive compensation and the length of the scheduling time on the willingness of the user group and finally synthesizes the above model that is used to construct the charging model of a single charging pile. Example proves that the scheduling strategy constructed in this chapter can effectively reduce the peak-valley difference of the system and also increase the revenue for EV aggregators and users, which significantly improves the social benefits.

Acknowledgements This work was financially supported by the State Grid Shanghai Electric Power Economic and Technology Research Institute Project (52090R23000W).

Nomenclature

Variables	Definition
t_d	Time the EV user participates in the scheduling
t_g	Time the EV is on the power grid
λ	The proportion of time that users are involved in scheduling
P_C	Maximum charging power of electric vehicles
P_D	Maximum discharge power of electric vehicles
η_c	Electric vehicle charging efficiency (kW)
η_d	Electric vehicle discharging efficiency (kW)
B	Electric vehicle battery capacity (kWh)
$cc \in \{0, 1\}$	Electric vehicle charging status, 1 indicates charging, 0 indicates not charging
$cd \in \{0, 1\}$	Electric vehicle discharging status, 1 indicates charging, 0 indicates not charging
ec	Maximum charging capacity of electric vehicles within a period of time
ed	Maximum discharging capacity of electric vehicles within a period of time
n_a	Power grid access period
$n_{\min}$	Minimum number of charging periods
$n_{\max}$	Maximum number of periods in power grid

n_{add}	Number of additional scheduling periods
n_{all}	Total number of scheduling periods
n_s	Schedulable time period

References

1. Huang, J., Luo, Z., Li, X., Zhou, R., et al.: Optimal scheduling of power systems with virtual power plant considering operation scenarios and their network loss. Power Syst. Technol. **15**, 1–8 (2025)
2. Koc, H., Gajjar, A., Graham, S.W.: Static charging scheduling of electric vehicles at private stations. In: Proceedings of the 2019 IEEE 10th Annual Information Technology, Electronics and Mobile Communication Conference (IEMCON), Vancouver, BC, Canada, pp. 0237–0240 (2019). https://doi.org/10.1109/IEMCON.2019.8936213
3. Mukherjee, J.C., Gupta, A.: A review of charge scheduling of electric vehicles in smart grid. IEEE Syst. J. **9**(4), 1541–1553 (2015). https://doi.org/10.1109/JSYST.2014.2356559
4. Shuai, T.J.F., Dongyang, Y., et al.: Accurate regulation method for virtual power plants considering static voltage load characteristics. Trans. China Electrotech. Soc. **24**, 1–18 (2025)
5. Sheik Mohammed, S., Imthias Ahamed, T.P., Devaraj, D.: Optimized charge scheduling of plug-in electric vehicles using modified placement algorithm. In: Proceedings of the 2019 International Conference on Computer Communication and Informatics (ICCCI), Coimbatore, India, pp. 1–5 (2019). https://doi.org/10.1109/ICCCI.2019.8821960.
6. Guangchao, D.: Research on Air conditioning Load Allocation and Scheduling Strategy for Multiple Refrigeration Plants Based on Load Forecasting. Zhejiang University (2024). https://doi.org/10.27461/d.cnki.gzjdx.2024.000623
7. Lei, M.: Research on Electric Vehicle Charging Load Forecasting and Optimizal Dispatching Strategy for Active Distribution Network Mao Lei. Shaanxi University of Science and Technology (2024)
8. Xu, J. Research on Scheduling Strategy of Flexible Load Participation Demand Response Based on Multi-Agent System. Hangzhou Dianzi University (2024)
9. Cheng, X.: Analysis of power load forecasting and scheduling strategy optimization based on machine learning. Electr. Technol. **53**(06), 326–327 (2024)
10. He, Z.: Single-Double Incentive Strategy for Low-Carbon Dispatch of Electric Vehicle Loads. Shanxi University (2024). https://doi.org/10.27284/d.cnki.gsxiu.2024.000348.
11. Yuan, H., Hu, G., Wang, X., et al.: General architecture of grid adjustable load resource scheduling control system. Yunnan Electr. Power **24**(04), 13–18 (2024)
12. Zhu, D., Jia, Y., Zhou, Q.: A load-source coordinated scheduling method based on resource flexibility of load aggregators. Zhejiang Electr. Power **43**(04), 12–20 (2024)
13. Yang, X., Qi, Q., Li, Q., et al.: Peak load game scheduling strategy combining reward mechanism and user willingness. Trans. China Electrotech. Soc. **39**(16), 5060–5074 (2024)
14. Su, Z.: Economic Operation of Wind-PV-ES Hybrid Microgrid by Considering of Coordination Between Flexible Load and Energy Storage System. Changsha University of Science and Technology (2022). https://doi.org/10.26985/d.cnki.gcsjc.2022.001057
15. Sun, Z., Liu, H., Liu, R., et al.: Optimal dispatch for microgrids with CHP considering EV users' intention. Electr. Power Constr. **39**(09), 95–104 (2018)
16. Yin, Q.: Research on the Regulation Strategy of Orderly Charging and Discharging of Electric Vehicles Considering Renewable Energy Access. Northeast Electric Power University (2023). https://doi.org/10.27008/d.cnki.gdbdc.2023.000164
17. He, X., Liao, S., Zhao, Z., et al.: The impact of large-scale development of urban electric vehicles on the atmospheric environment. Chin. J. Environ. Manag. **12**(03), 77–84 (2020)

18. Li, P., Ma, X., Guo, J., et al.: Game strategy of integrated energy service providers based on bi-level optimization. Power Syst. Technol. **45**(02), 460–473 (2021)
19. Xie, H.: Research on charge and discharge optimization strategy of electric vehicle considering battery loss. Xiamen Univ Technol **54**, 4552 (2023)

Multisource Information Fusion-Based Bus Origin–Destination Prediction Model

Zeyu Wen, Mingquan Shi, Yi Li, and Wenyuan Wu

Abstract Efficient management and accurate prediction of urban traffic flow have become critical challenges in modern smart city governance, where origin–destination (OD) flow prediction plays a vital role in optimizing traffic resource allocation. Existing research indicates that the type distribution, density, and spatial distance of points of interest (POIs) surrounding stations are significant factors influencing passenger travel decisions. However, traditional prediction models, when characterizing the complex associations between POI features and passenger travel purposes, often overlook the dynamic interactive relationship between passenger types and travel intentions, resulting in predictions that inaccurately reflect actual travel behavior patterns. To address this issue, this chapter proposes a bus OD prediction framework based on multisource information fusion. The model comprises three key modules: a feature extraction module based on passenger types and station POIs, a time-series prediction module leveraging historical OD data, and an innovative conservative fusion mechanism. Its key innovations and advantages include the following: firstly, the fine-grained modeling of travel preferences for diverse passenger groups (e.g., commuters, students, and tourists) through a parameterized interaction matrix between passenger types and POIs; secondly, the employment of a multi-head self-attention mechanism for deep feature extraction from spatiotemporal data, enhancing the model's capacity to capture temporal patterns and long-term dependencies; and, finally, a conservative fusion strategy dynamically allocates model weights and introduces residual connections, effectively integrating prior knowledge from POIs with statistical patterns from historical data, while also incorporating multidimensional external factors such as weather conditions and holiday information, thereby constructing a robust and highly interpretable comprehensive prediction framework. Experimental results demonstrate that the proposed model significantly outperforms traditional methods in multiday prediction tasks, exhibiting

Z. Wen
School of Computer Science and Technology, Chongqing University of Posts and Telecommunications, Chongqing, China

M. Shi (✉) · Y. Li · W. Wu
Chongqing Institute of Green and Intelligent Technology, Chinese Academy of Sciences, Chongqing, China
e-mail: shimq@cigit.ac.cn

K. Subramaniyam and M. Paprzycki (eds.), *Smart Cities and the Internet of Things*, Internet of Things, https://doi.org/10.1007/978-3-032-08290-9_12

superior adaptability, particularly under complex weather conditions and on special dates.

Keywords OD matrix · POI · Attention mechanism · Passenger type · Multisource information fusion · Transformer

1 Introduction

Origin–destination (OD) data are fundamental for analyzing passenger mobility patterns within public transport, a vital pillar of urban transit, offering significant value for studying urban passenger flow and travel behavior.

OD prediction technology has evolved from traditional statistical methods (e.g., autoregressive integrated moving average [ARIMA] [1]) and machine learning approaches (e.g., support vector machines and random forests [2, 3]), which showed limitations in capturing complex nonlinear spatiotemporal relationships, to modern deep learning models. Deep learning models, leveraging powerful feature extraction capabilities and high-precision performance, have gradually become the mainstream research direction. For instance, deep learning models have incorporated clustering [4], unified spatiotemporal representations, and graph neural networks [5, 6] to capture spatiotemporal dependencies, while hybrid frameworks [7] balanced efficiency and accuracy.

Further optimizations include exploring correlations with points of interest (POIs) [8], considering external factors [9], employing generative adversarial networks, and using multimodel ensembles[10].

While existing studies acknowledge POIs' influence on alighting choices, they often lack fine-grained modeling of the complex POI–passenger behavior association. Notably, different passenger types exhibit distinct preferences for POI categories. To address this, we propose a multisource fusion framework[11] featuring: (1) an OD prediction module with a parameterized alighting intention matrix quantifying passenger type-POI preferences and (2) a transformer-based module for historical OD pattern extraction[12]. These are integrated via a conservative fusion mechanism, also incorporating external factors (weather and holidays).

2 Problem Analysis

2.1 Problem Formulation

This chapter addresses the problem of regional passenger flow prediction in cities. The main parameters and their meanings are shown in Table 1.

Table 1 List of major symbols and descriptions

Symbol	Dimension	Description
V	$\mathbb{R}$	Number of regional stations
n	$\mathbb{R}$	Number of passenger IC card types
m	$\mathbb{R}$	Number of POI types
EF	$\mathbb{R}$	External factors including time and weather
U	$\mathbb{R}^{v\times n}$	Distribution of boarding passenger types at regional stations
O	$\mathbb{R}^{v\times v}$	Regional OD passenger flow matrix
A	$\mathbb{R}^{v\times v}$	Adjacency matrix of stations

All stations within a region are connected according to actual routes; different routes may share common stations, forming the regional public transport network through these hub stations. Here, $V = (v_1, \ldots, v_n)$ represents the set of regional public transport stations, and A denotes the regional station adjacency matrix. If station v_i is connected to station v_j $(i, j \in v)$, then $A_{ij} = 1$; otherwise, $A_{ij} = 0$.

Based on the concepts introduced above, the OD passenger flow prediction problem is defined as follows:

Input: Historical OD passenger flow $O^{t-T+1:t}$, the historical distribution matrix of boarding passenger types $U^{t-T+1:t}$, future OD passenger flow $O^{(t+1:t+H)}$, the distribution of boarding passenger types at regional stations U^f, and external influencing factors EF.

Output: A model M satisfying

$$\hat{O}^{(t+1:t+H)} = \text{model}\left(O^{t-T+1:t}, U^{t-T+1:t}, \text{EF}\right)$$

Objective:

$$\text{Min}\left(\hat{O}^{(t+1:t+H)} - O^{(t+1:t+H)}\right)$$

2.2 Analysis of Influencing Factors

Public transport OD flow is influenced by temporal, passenger type, and spatial characteristics. Our framework uses a transformer-based module to capture diverse temporal patterns (e.g., weekly, seasonal, and weather-related).

Different passenger types exhibit distinct preferences for POI categories (e.g., educational, commercial, and leisure/medical [13, 14]), with external factors also having differential impacts.

Thus, we introduce a learnable alighting intention matrix for fine-grained, interpretable mapping of passenger type preferences to POIs, improving OD prediction.

Table 2 POI types

Financial and insurance services	Auto repair	Auto sales
Shopping services	Catering services	Transportation facility services
Life services	Scenic spots	Sports and leisure services
Science, education and cultural services	Medical and healthcare services	Companies and enterprises
Commercial housing	Accommodation services	

3 Prediction Model Construction

3.1 Passenger Flow Prediction Module Based on Passenger Type and POI

3.1.1 POI Setup and External Factor Construction

We selected 14 influential POI types relevant to public transport travel from the Amap (Gaode Maps) classification system (Table 2), covering diverse activity locations. A distance-based linear decay weight function quantifies POI impact: weight decreases from 1 to 0.125 for POIs within an 800 m radius of a station.

Temporal features (e.g., weekday/weekend and holiday status) and weather features (e.g., type, intensity, and numerically encoded) were integrated as external factors[15].

3.1.2 Model Construction

This module predicts OD distribution by modeling passenger type–POI associations via three submodules, as illustrated in Fig. 1.

Passenger Distribution Prediction Submodule:

This submodule first constructs a time-series input sequence (X, Eq. 1) from historical passenger type distribution $U^{(t-T+1:t)}$ and weather features ($W \in \mathbb{R}3$). Subsequently, a transformer encoder (Eq. 2) extracts deep temporal features (H), which are then fed into a transformer decoder to predict the passenger type distribution for future H time steps ($U^{(t+1:t+H)}$; Eq. 3).

$$X = \left[U_{t-N+1:t}; W_{t-N+1:t}\right] \in \mathbb{R}^{N\times(v\times n+3)} \tag{1}$$

$$H = \text{Transformer Encoder}(X) \in \mathbb{R}^{N\times d_{\text{model}}} \tag{2}$$

$$U_{t+1:t+H} = \text{Transformer Decoder}(H) \in \mathbb{R}^{H\times v\times n} \tag{3}$$

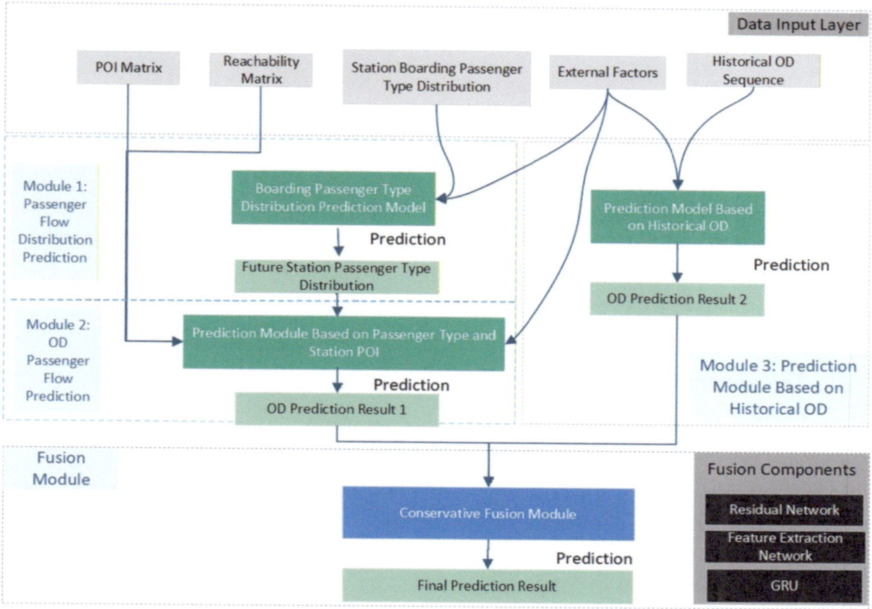

Fig. 1 Multisource information fusion OD prediction framework

POI Association Modeling Submodule:

Central to this submodule is the introduction of a learnable alighting intention matrix ($D \in \mathbb{R}^{n\times m}$), parameterizing the preference strength of different passenger types for various POI categories. This matrix, along with the predicted passenger distribution, station POI feature matrix ($P \in \mathbb{R}^{v\times m}$), is used to compute an initial OD distribution. This is further refined through a fully connected layer for enhanced feature representation (O_{init}; Eq. 5).

$$B = U \cdot D, C = B \cdot P^T \tag{4}$$

$$O_{\text{init}} = \text{ReLU}(W_2 \cdot \text{concat}(\text{embedding}(U), P)) \tag{5}$$

Constraint Optimization Submodule:

To ensure the physical plausibility and stability of predictions, multiple constraints are incorporated: a nonnegativity constraint enforced by a rectified linear unit (ReLU) activation function (Eq. 6) and a network topology constraint utilizing the station adjacency matrix A (Eq. 7). The model is optimized using a composite loss function (Eq. 8) that includes mean squared error (MSE; L_{MSE}), a temporal smoothing loss (L_{smooth}), and a regularization term ($\|W1\|_2$).

$$O_{\text{pred}} = \text{ReLU}(O_{\text{init}}) \tag{6}$$

$$O_{\text{final}} = O_{\text{pred}} \odot A \tag{7}$$

$$L = \alpha L_{\text{MSE}}(O_{\text{final}}, O_{\text{true}}) + \beta L_{\text{smooth}} + \gamma \|W1\|_2 \tag{8}$$

Through parameterized learning, particularly the design of the alighting intention matrix, this module achieves a fine-grained mapping between passenger travel decisions and urban functional layout. Coupled with multilevel constraints, it ensures both prediction accuracy and interpretability.

3.2 *Transformer Prediction Model Based on Historical OD Data*

This module is a deep learning model based on the transformer architecture, used to predict future origin–destination (OD) matrices. By analyzing historical OD data and weather features, the model captures long-term dependencies in the time series, thus enabling prediction of future OD matrices. The model receives two types of input: One is the historical OD sequence O^h, and the other is the weather feature sequence EF.

Here, B is the batch size, T is the input time window length, and N is the number of valid station pairs.

First, feature embedding is performed. The model concatenates OD data and weather features and transforms their dimensions.

$$X_{\text{combined}} = \text{Concat}(X_{\text{OD}}, X_W) \tag{9}$$

Next, feature embedding is done through a linear transformation.

$$E = W_e X_{\text{combined}} + b_e \tag{10}$$

Here, W_e and b_e are learnable parameter matrices and bias terms.

The embedded sequence is then processed by multilayer transformer encoders to capture temporal dependencies. The output from the final time step of these encoders (H) is subsequently fed into a two-layer fully connected network to generate the predicted OD matrix (Y).

$$H = \text{ReLU}(\text{Layer Norm}(W_1 E_{\text{last}} + b_1)) \tag{11}$$

$$Y = W_2 H + b_2 \tag{12}$$

This module effectively utilizes sequence information, captures dependencies between different time steps through the self-attention mechanism, integrates weather impact by incorporating weather features as auxiliary information, supports multiday prediction, reduces computational complexity by predicting only valid station pairs through dimensionality reduction, and finally uses mean squared error as the loss function for the training objective.

3.3 Conservative Fusion Mechanism

This section employs a conservative fusion mechanism based on residual learning to merge OD matrix prediction results from different sources. The architecture of this module is detailed in Fig. 2.This mechanism uses a dynamic weight allocation strategy, while ensuring the reliability of prediction results through residual transformation and multiple loss constraints.

3.3.1 Overview of the Fusion Mechanism

Let $\mathrm{OD}_1 \in \mathbb{R}^{B\times M\times N}$ be the OD matrix generated by the passenger flow prediction model processed by the feature extraction network (secondary prediction source), and $\mathrm{OD}_2 \in \mathbb{R}^{B\times M\times N}$ be the OD matrix generated by the historical data prediction model (primary prediction source). Here, B is the batch size, M is the number of prediction days, and N is the number of valid station pairs. The fusion process includes the following key components:

First is the feature extraction network, which is implemented by a multilayer perceptron, including linear transformation, layer normalization, ReLU activation, and dropout regularization.

$$F_1 = \text{Feature Extract}(\mathrm{OD}_1) \tag{13}$$

$$F_2 = \text{Feature Extract}(\mathrm{OD}_2) \tag{14}$$

Second is the residual transformation network, which only performs residual transformation on the secondary prediction source:

$$R = \text{Residual Transform}(\mathrm{OD}_1) \tag{15}$$

Third is the gating network, which calculates fusion weights based on the features from the two prediction sources:

$$\alpha = \sigma(W_2\mathrm{ReLU}(W1[F_1; F_2] + b_1) + b_2) \cdot \gamma \tag{16}$$

Here, γ is a scaling factor. The sigmoid activation function σ ensures base weight values are within [0,1], and γ further modulates the influence of the secondary source.

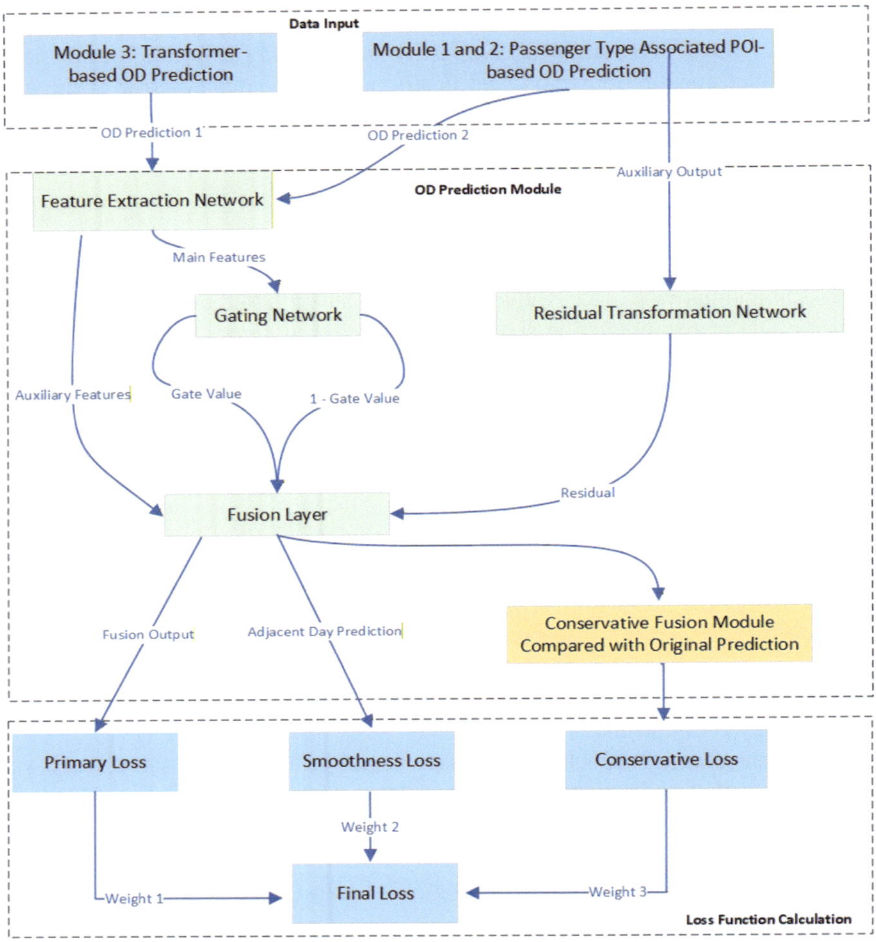

Fig. 2 Conservative fusion module

The final fusion result is obtained through a weighted combination:

$$OD_{pred} = OD_2 \cdot (1 - \alpha) + R \cdot \alpha \tag{17}$$

where α is the gating value.

This design ensures that the fusion result primarily depends on historical data prediction (OD_2). This prioritization is based on the established principle that historical OD patterns, due to their inherent regularity and periodicity, provide a robust and reliable baseline for future demand forecasting. The passenger flow prediction (via R) then offers a data-driven refinement to this baseline.

To ensure the quality of the fusion result, a triple loss function is designed: First, the main loss ensures prediction accuracy. Second, the smoothness loss guarantees the smoothness of predictions across adjacent days. Third, the conservative loss prevents predictions from deviating excessively from the primary prediction source:

$$\begin{aligned} L_{\text{main}} &= \text{MSE}(\text{OD}, Y)\, L_{\text{smooth}} = \text{MSE}(\text{OD}_{t+1}, O_t)\, L_{\text{conservative}} \\ &= \text{MSE}\big(\text{OD}_{\text{pred}}, \text{OD}_2\big) \end{aligned} \tag{18}$$

$$L_{\text{total}} = L_{\text{main}} + aL_{\text{smooth}} + bL_{\text{conservative}} \tag{19}$$

The coefficients a and b in Eq. (19) balance the contributions of the smoothness and conservative loss terms, respectively. These, along with the scaling factor γ from Eq. (16), are hyperparameters whose values are optimized during model training, as detailed in Sect. 4.2. This multiloss design effectively balances the three objectives of prediction accuracy, temporal smoothness, and prediction conservativeness.

4 Experiments

4.1 Research Data

This study selected public bus operation data from Chongqing City for the entire year of 2023 (January 1 to December 31), using the day as the basic time granularity, aiming to predict future passenger travel demand. The dataset covers multidimensional data including passenger card swipe records, bus operation data, and station geographical location information. The experiment focuses on regional origin–destination (OD) data comprising 10 bus routes and 96 stations within the Caijia area of the city.

The study area in Caijia, Chongqing, characterized by its high-density station layout, diverse route types, and complex urban functions, provides an ideal and challenging environment for model validation.

4.2 Model Parameter Settings

The key parameter configurations used in this study are as follows: input time window $n = 10$ days and prediction time step $m = 3$ days. The transformer module for passenger count prediction uses 3 attention heads, a hidden layer dimension of 124, 3 submodules, and a dropout rate set to 0.1. The OD prediction component based on passenger travel volume uses an encoding layer dimension of 512 and a dropout rate of 0.1. The prediction module based on historical OD data employs a 3-layer transformer structure with 4 attention heads, an encoding hidden layer dimension of 124, and a feed-forward network hidden layer dimension of 2048, building deep spatiotemporal feature extraction capabilities.

The alighting intention matrix D previously defined as $D \in \mathbb{R}^{n \times m}$ (see Table 1 for symbol definitions). For this study, $n = 9$ passenger types, based on predefined Chongqing Public Transport Corporation categories, and $m = 14$ POI types, selected from Amap (Gaode Maps) for relevance to public transport (details in Table 2).

Key hyperparameters for the fusion mechanism (Eqs. 16–19) were optimized on a validation set. The scaling factor was set to $\gamma = 0.1$. Loss weights were $a = 0.1$ for L_{smooth} and $b = 0.01$ for $L_{\text{conservative}}$, balancing accuracy, smoothness, and conservativeness.

4.3 Experiments and Discussion

To comprehensively evaluate the performance of the multisource information fusion model (MIFM) and validate the effectiveness of the user–POI association mechanism, the following ablation experiments were designed:

(1) Complete multisource information fusion model (MIFM)
(2) Transformer prediction model based solely on historical OD data (TPMOD)
(3) Passenger flow prediction module based solely on passenger type and POI (PFPM)

To comprehensively evaluate the performance of the MIFM model, this study selected five representative baseline models for comparison: graph convolutional network (GCN), diffusion convolutional recurrent neural network (DCRNN), spatiotemporal attention graph convolutional network (ASTGCN), long short-term memory (LSTM), and spatiotemporal attention network (STAN). Table 3 shows the detailed performance of each model across the three prediction steps.

MIFM significantly outperformed all baseline models (Table 3). Compared to the best baseline (STAN), MIFM reduced MSE by approximately 18.6% on average. It also demonstrated superior mean absolute error (MAE; stable ~ 2.16–2.19) and prediction stability, with its MSE increasing by only 3.3% over three days, far lower than GCN (4.7%) and DCRNN (14.3%). These results highlight MIFM's advantages in accuracy and stability, especially for multiday forecasts.

Table 3 Results from the ablation experiments

Model	Day 1		Day 2		Day 3	
	MSE	MAE	MSE	MAE	MSE	MAE
MIFM	30.462	2.163	31.015	2.174	31.480	2.187
TPMOD	31.343	2.172	32.117	2.182	32.362	2.197
PFPM	59.887	2.282	59.949	3.066	60.088	3.060

Table 4 Performance analysis of the comparative experiments

Model	Step 1		Step 2		Step 3	
	MSE	MAE	MSE	MAE	MSE	MAE
GCN	101.28	4.6	103.33	4.61	106.07	4.71
DCRNN	93.73	4.42	94.7	4.29	107.15	4.55
ASTGCN	130.71	2.28	132.69	2.28	131.42	2.28
LSTM	40.39	2.16	42.99	2.22	44.94	2.2
STAN	38.7	2.65	36.87	2.58	36.79	2.56
MIFM	30.4621	2.1629	31.0148	2.1748	31.48	2.1867

4.4 Comparison with Baseline Models

To comprehensively evaluate the performance of the MIFM model, this study selected five representative baseline models for comparison: graph convolutional network (GCN), diffusion convolutional recurrent neural network (DCRNN), spatiotemporal attention graph convolutional network (ASTGCN), long short-term memory (LSTM), and spatiotemporal attention network (STAN). Table 4 shows the detailed performance of each model across the three prediction steps.

MIFM significantly outperformed all baseline models (Table 3). Compared to the best baseline (STAN), MIFM reduced MSE by approximately 18.6% on average. It also demonstrated superior MAE (stable ~ 2.16–2.19) and prediction stability, with its MSE increasing by only 3.3% over three days, far lower than GCN (4.7%) and DCRNN (14.3%). These results highlight MIFM's advantages in accuracy and stability, especially for multiday forecasts.

5 Conclusion

This chapter introduced a multisource information fusion model (MIFM) for urban bus origin–destination (OD) prediction, effectively addressing the challenge of capturing complex passenger–POI interactions and spatiotemporal dependencies. MIFM uniquely combines a passenger flow prediction module with a learnable alighting intention matrix, a transformer for historical OD patterns, and a conservative fusion mechanism that incorporates external factors.

Experiments on real-world Chongqing bus data demonstrated MIFM's significant superiority. It outperformed strong baselines (e.g., ~ 18.6% MSE reduction versus STAN) and showed robust multiday forecasting stability. Ablation studies confirmed the critical contributions of the passenger type–POI association and the multisource fusion approach.

The primary contributions are the novel fine-grained modeling of passenger type–POI preferences, enhancing interpretability, and an effective fusion strategy that improves prediction accuracy and reliability. Future work may focus on dynamic passenger/POI modeling, real-time data integration, and broader applicability. In summary, MIFM offers a promising and robust solution for OD prediction, valuable for optimizing urban public transit systems.

References

1. Miloš, M., Libor, Š, Vlastimil, M., et al.: SARIMA modelling approach for railway passenger flow forecasting. Transport **33**(5), 1113–1120 (2018)
2. Rodriguez-Rueda, P.J., Ruiz-Aguilar, J.J., Gonzalez-Enrique, J., et al.: Origin-destination matrix estimation and prediction from socioeconomic variables using automatic feature selection procedure-based machine learning model. J. Urban Plan. Dev. **147**(4), 04021056 (2021)
3. Lagos, F., Moreno, S., Yushimito, W.F., et al.: Urban origin-destination travel time estimation using K-nearest-neighbor-based methods. Mathematics **12**(8), 1255 (2024)
4. Ke, J., Zheng, H., Yang, H., et al.: Short-term forecasting of passenger demand under on-demand ride services: a spatio-temporal deep learning approach. Transp. Res. Part C Emerg. Technol. **85**, 591–608 (2017)
5. Zhao, L., Song, Y., Zhang, C., et al.: T-GCN: a temporal graph convolutional network for traffic prediction. IEEE Trans. Intell. Transp. Syst. **21**(9), 3848–3858 (2020)
6. Guo, S., Lin, Y., Feng, N., et al.: Attention based spatial–temporal graph convolutional networks for traffic flow forecasting. Proc. AAAI Conf. Artif. Intell. **33**(1), 922–929 (2019)
7. Wu, Y., Tan, H., Qin, L., et al.: A hybrid deep learning based traffic flow prediction method and its understanding. Transp. Res. Part C Emerg. Technol. **90**, 166–180 (2018)
8. Zeng, W., Fu, C.W., Arisona, S.M., et al.: Visualizing mobility of public transportation system. IEEE Trans. Visual. Comput. Graphics **20**(12), 1833–1842 (2014)
9. Klar, R., Rubensson, I.: Spatio-temporal investigation of public transport demand using smart card data. Appl. Spat. Anal. Policy **17**(1), 241–268 (2024)
10. Liu, L., Zhu, Y., Li, G., et al.: Online metro origin-destination prediction via heterogeneous information aggregation. IEEE Trans. Pattern. Anal. Mach. Intell. **45**(3), 3574–3589 (2023)
11. Li, J., Wang, Y., Zhang, T., et al.: Multi-source fusion with graph neural networks for urban mobility prediction. In: Proceedings of the 31st ACM International Conference on Information & Knowledge Management, pp. 1234–1243 (2023)
12. Chen, X., Liu, Z., Sun, H., et al.: Dynamic OD prediction in public transport: a transformer-based approach with spatio-temporal attention. IEEE Trans. Intell. Transp. Syst. **25**(2), 1456–1468 (2024)
13. Zhao, X., Cui, M.Y., Levinson, D.: Exploring temporal variability in travel patterns on public transit using big smart card data. Environ. Plan. B Urban Anal. City Sci. **50**(1), 198–217 (2023)
14. Liu, Y.Z., Cheng, T.: Understanding public transit patterns with open geodemographics to facilitate public transport planning. Transp. Transp. Sci. **16**(1), 76–103 (2020)
15. Huang, Q.Y., Yang, Y.J., Xu, Y.B., et al.: Human origin-destination flow prediction based on large scale mobile signal data. Wirel. Commun. Mob. Comput. **1**, 1604268 (2021)

Application and Empowerment Path of Digital Twin Technology in Automobile Manufacturing Industry

Yongkang Liang, Kuo Song, Shun Wang, and Jingli Han

Abstract Digital twin technology is widely used in automobile design, production, maintenance, and other stages, providing technical support for the transformation and upgrading of automobile manufacturing industry. However, problems such as complex technology integration, data security issues, and shortage of professionals are also prominent. To this end, the automobile manufacturing industry can be empowered by developing an integrated design platform, establishing a network security system, and strengthening the training of technical personnel.

Keywords Digital twin technology · Automobile manufacturing industry · Digital transformation

1 Introduction

In today's era of rapid development of science and technology, the automobile manufacturing industry is in the wave of profound changes [1]. Digital transformation has become the only way for the development of many industries, including the automobile manufacturing industry [2]. In the process of digitization, digital twin technology, as a very forward-looking and transformative emerging technology, has gradually entered the field of vision of the automobile manufacturing industry and is expected to become a key force to promote its transformation and upgrading.

Some scholars have discussed the application of digital twin technology in automobile manufacturing workshop. Liu et al. [3] introduced digital twin technology into the automotive remanufacturing workshop to promote the recycling of resources; Leng et al. [4] successfully solved the compatibility problem of production line equipment by testing the remote and near-physical debugging method of digital twin technology on the smartphone production line; Li et al. [5] developed the digital twin modeling technology of the body-in-white welding production line, which effectively reduced the time required for production line construction and debugging;

Y. Liang (✉) · K. Song · S. Wang · J. Han
School of Economics and Management, Yan'an University, Yan'an, Shaanxi, China
e-mail: ykliang85@163.com

K. Subramaniyam and M. Paprzycki (eds.), *Smart Cities and the Internet of Things*, Internet of Things, https://doi.org/10.1007/978-3-032-08290-9_13

Yan et al. [6] used digital twin technology to shorten the design cycle in the personalized customized design of panel furniture production line. Son et al. [7] developed a cyber-physical system (CPS) based on digital twin technology to predict the production capacity of the body production line and verify the feasibility of production scheduling. Although its theoretical advantages are obvious, but in the process of integration with the complex industrial system of automobile manufacturing, it faces many difficulties, which seriously restricts the wide application and in-depth promotion of this technology in the industry. Therefore, it is extremely urgent and important practical significance to carry out in-depth research on the predicament and path of the development of the automobile manufacturing industry.

2 Current Status of Development of the Automobile Manufacturing Industry

In recent years, the overall scale of China's manufacturing industry has ranked first in the world for many consecutive years, and the automobile manufacturing industry, as an important part of it, has a considerable scale (Fig. 1). According to statistics, in 2023, auto production and sales exceeded 30 million for the first time, achieving 30.161 million and 30.94 million, respectively, an increase of 11.6 and 12% year on year, reaching a record high. It is predicted that the annual automobile production and sales will reach 31 million in 2024, an increase of more than 3%, and the industrial scale will continue to rise.

With the increasing attention of consumers to environmental protection and energy conservation and the increasing maturity of new energy vehicle technology, the market demand of traditional fuel vehicles has gradually declined. With the advantages of environmental protection and energy conservation, the market share of new energy vehicles has been continuously improved and the industry has developed rapidly (Fig. 2). In 2023, the production and sales of new energy vehicles have

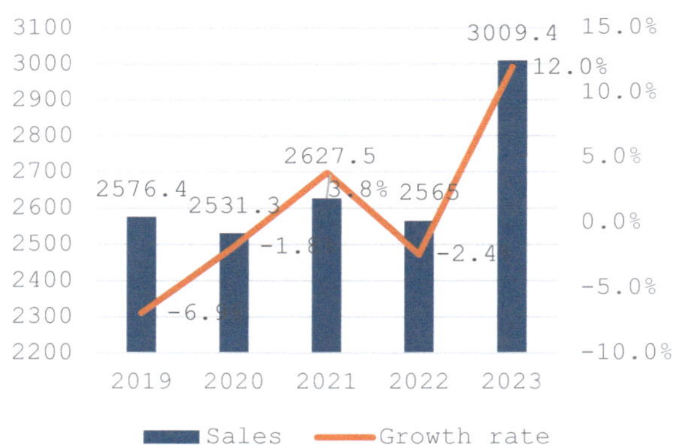

Fig. 1 Annual sales and growth rate of China's automobiles in the past five years. Data sources: China Association of Automobile Manufacturers (CAAM) Information Conference in the past five years

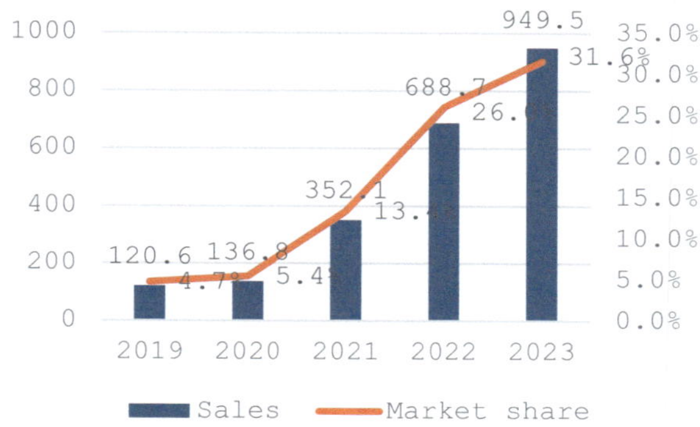

Fig. 2 New energy vehicle market size in the last five years. Data sources: China Association of Automobile Manufacturers (CAAM) Information Conference in the past five years

completed 9.587 million and 9.495 million, respectively, an increase of 35.8% and 37.9%, respectively, and the market share has been steadily improved. It is expected that China's new energy vehicle production and sales will reach about 11.5 million in 2024, an increase of about 20%.

3 Digital Twin Application Scenarios

3.1 Automobile Design Stage

By establishing a high-precision digital twin model, designers can boldly try new design concepts and functions in a virtual environment (Fig. 3). Digital twin technology can be used to simulate the aerodynamics of automobiles more accurately. Through computational fluid dynamics simulation technology, designers can optimize the aerodynamic design of vehicles to the maximum extent in the digital twin model, reduce the wind resistance coefficient, and improve the energy efficiency of automobiles. In addition, the electric vehicle battery can be coupled and abstracted into a design model and a vehicle simulator through digital twin technology, so as to control the closed-loop real-time vehicle system based on the digital twin of the battery.

3.2 Automobile Production Stage

The application of digital twin technology optimizes the automobile production process. A large number of sensors are deployed on the production line to collect data such as equipment operation status, production process parameters, and material flow in real time. Using data analysis and simulation technology, the production process is modeled and simulated to find out the bottleneck links and optimization points

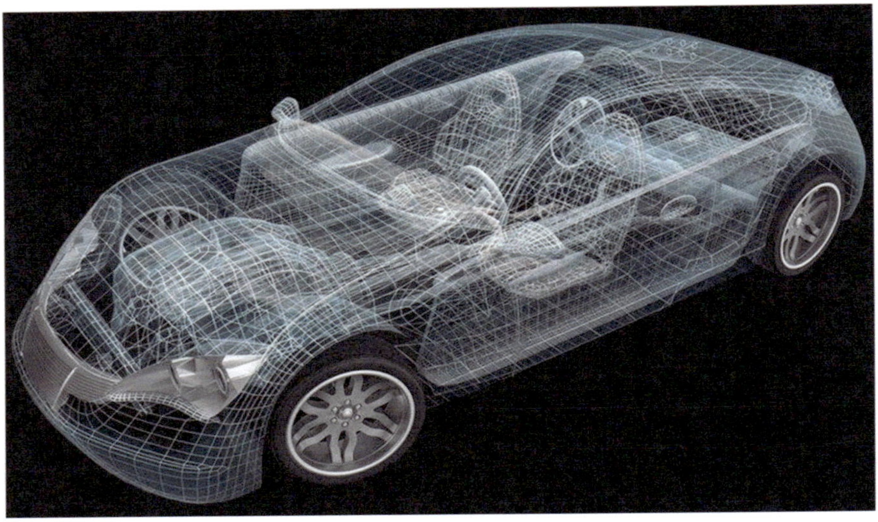

Fig. 3 Automobile simulation design and assembly VR system. Source of information: http://sufencg.com/a/chanpinzhanshi/chanpinfenleimingchenDIY/2018/0228/338.html

in the production process. Then adjust the production plan; optimize the process parameters; improve the equipment layout, etc.; improve production efficiency and product quality; and reduce production costs. Figure 4 shows the local part of Lotus global intelligent factory.

3.3 *Car Maintenance Phase*

With the help of on-board sensors, the vehicle operation data are collected in real time, and the digital twin model of the vehicle is constructed to realize health monitoring and fault prediction (Fig. 5). By comparing the analysis data with the normal state, potential faults can be detected in advance, such as predicting battery decay, timely warning, and reducing driving risk. In terms of remote diagnosis and maintenance, when the fault occurs, the vehicle data are transmitted to the platform, and the maintenance personnel can remotely analyze the twin model to determine the fault point, formulate the maintenance plan, and obtain remote technical support to improve the maintenance efficiency.

Fig. 4 Local schematic diagram of Lotus global intelligent factory. Source of information: https://zhuanlan.zhihu.com/p/639295542

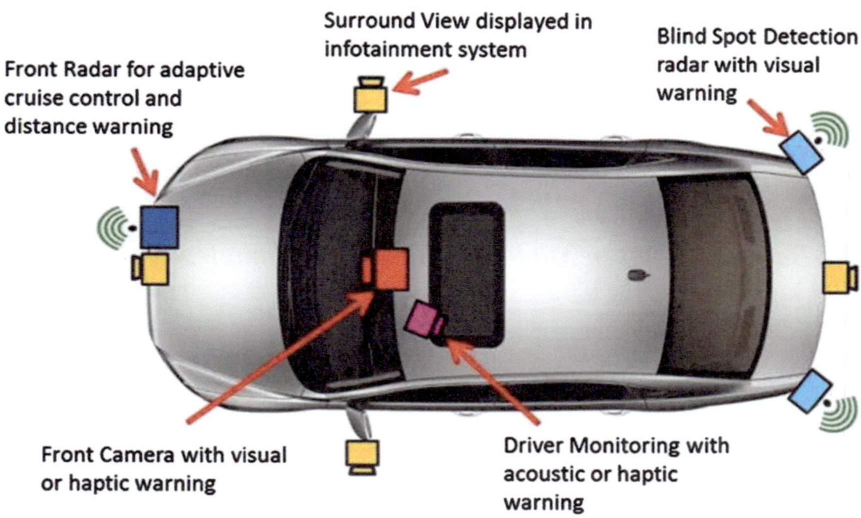

Fig. 5 Vehicle sensor demonstration. Source of information: http://www.uml.org.cn/embeded/201903291.asp

4 Challenges of the Application of Digital Twin Technology in Automobile Manufacturing Industry

4.1 Complexity of Technology Integration

The application of digital twin technology in automobile manufacturing industry involves knowledge and technology in many fields, and its integration process faces many difficulties [8]. Digital twin technology covers many key technical fields such as modeling and simulation, data acquisition and processing, Internet of Things, visualization, and interaction technology. In terms of modeling and simulation, it is necessary to accurately construct the digital model of the automobile and its production process. It is required to have a deep understanding of the geometric structure of the automobile and master complex technologies such as physical modeling and behavior modeling, in order to accurately simulate the performance of the automobiles under different working conditions.

4.2 Data Security and Privacy Protection

In the automobile manufacturing industry, data security is very important [9], and automobile manufacturers are facing network security risks [10]. A large amount of data is generated in the process of automobile manufacturing, including the business secrets of enterprises and the personal privacy of users. Data leakage or malicious use will bring huge losses to enterprises and users.

4.3 Balance Between Model Accuracy and Complexity

When constructing the automobile digital twin model, it is necessary to balance between accuracy and complexity. High-precision models can more accurately simulate the physical characteristics and behavior of vehicles, but they are often accompanied by high complexity. For example, when establishing the digital twin model of automobile collision, in order to accurately simulate the deformation of the body and the damage of parts, it is necessary to consider the nonlinear mechanical properties of materials, the mechanical properties of solder joints, and other complex factors, which will make the model very complicated. When a model is too complex, it will increase the computational cost and time. In practical applications, models need to be able to simulate and analyze quickly to support automotive design and production decisions. For example, in the optimization process of automobile production lines, it is necessary to frequently use the digital twin model to simulate different production scenarios. If the model calculation speed is too slow, it cannot meet the actual needs.

4.4 *Model Update and Maintenance Costs Are High*

Automobile manufacturing technology is constantly developing, and products are constantly upgrading. When the design, process, or parts of the car change, the digital twin model needs to be updated in time. For example, when an automobile manufacturer launches a new electric vehicle, its battery system and power system are very different from traditional fuel vehicles, and the digital twin model needs to be reconstructed or modified. During the use of the car, the parts will gradually wear and age, which also needs to be reflected in the digital twin model. However, real-time monitoring and updating models to reflect these changes is a huge challenge and costly. For example, to update the digital twin model of automobile tire wear in real time, it is necessary to install a high-precision tire wear sensor and establish a corresponding data processing and model update mechanism.

4.5 *Demand for Professional Talents*

In the design stage, digital twin technology should be used by engineers to carry out innovative design to optimize the performance and reliability of automobiles. In the production stage, digital twin technology should be used by engineers to monitor the production process in real time and improve production efficiency and product quality. In the maintenance and after-sales stage, digital twin technology should be used by engineers for fault diagnosis and maintenance plan optimization.

5 Paths of Digital Twin Technology Enabling the Development of Automotive Manufacturing Industry

5.1 *Developing an Integrated Design Platform*

We integrate the multi-technology tool chain of IoT, big data, modeling and simulation, and artificial intelligence involved in digital twin technology. Engineers in automotive manufacturing companies can complete a series of work from the configuration of data acquisition equipment, creation and updating of digital models, to model analysis and optimization based on big data and artificial intelligence in a unified interface and environment, avoiding the complexity and inefficiency caused by switching between multiple different tools and platforms; we provide mechanisms for data sharing and interaction between tool chains to ensure that data can be accurately transferred and effectively utilized across tool chains.

5.2 Establishment of a Network Security System

Mature cloud platform security protection technologies such as firewalls, website application-level intrusion prevention systems, cloud situational awareness, database encryption, and database desensitization are adopted to ensure cloud data security; on the software side, the mobile app is security-hardened, combining security penetration testing, compliance testing, and security monitoring to eliminate app threats; on the vehicle side, based on the public key infrastructure (PKI) system and other infrastructure, we adopt security protection strategies such as deep defense for the whole vehicle, combined with user authorization and consent, authority control, security monitoring, and active defense technologies to achieve overall security protection on the vehicle side.

5.3 Model Simplification and Optimization Techniques Are Used

We use model reduction techniques to reduce the complexity of the digital twin model. For example, in the vehicle dynamics model, through the modal analysis of the vehicle system, the main modal information is extracted, and the secondary modes are ignored, thereby simplifying the model. At the same time, the surrogate model technology can be used to approximate the complex physical model with a simple mathematical model to improve the computational efficiency of the model. We establish a feedback mechanism for model optimization. According to the application scenarios and requirements of the model, the accuracy and complexity of the model are continuously adjusted. For example, in the early design stage of a car, a model with slightly lower accuracy but faster calculation speed can be used for conceptual design. In the detailed design stage, a high-precision model can be used for performance optimization.

5.4 Automatic Model Update Mechanism

By using the Internet of Things technology and big data analysis, the changes of automobile manufacturing process and vehicle running state are monitored in real time. For example, by installing IoT sensors on the automobile production line, the operating status of the equipment and the changes of process parameters are obtained in real time. When these parameters exceed a certain threshold, the update of the digital twin model is automatically triggered. Machine learning algorithms are used to predict the wear and aging of auto parts, and the digital twin model is updated in advance. For example, through the analysis of vehicle historical operation data and component life data, a component wear prediction model is established. When it is predicted that the component is about to reach the wear limit, the component state in the digital twin model is automatically updated.

5.5 Strengthening Technical Manpower Development

Automobile manufacturing enterprises should increase the training of employees, establish a talent incentive mechanism, encourage employees to learn and apply digital twin technology, and strengthen cooperation with universities and research institutions. Relevant industry associations in the automobile manufacturing industry should organize digital twin technology exchange activities, establish a digital twin technology talent certification system, standardize the talent market, improve the quality and level of talent, and improve the market competitiveness of talent.

6 Conclusions

Through the application of production line optimization, quality control, equipment maintenance, and production collaboration, digital twin technology can significantly improve the production efficiency and product quality of automotive manufacturing. With the continuous development and innovation of technology, digital twin technology will be deeply integrated with artificial intelligence, big data, cloud platforms, and other technologies, continuously expanding its application areas and functions and providing new impetus for the digital transformation and sustainable development of the automotive manufacturing industry.

Acknowledgements This work was supported by the Youth Fund of the Ministry of Education for Humanities and Social Sciences Research (23YJC630102) and The Doctoral Project of Yan'an University (YDBK2020-08).

References

1. Liu, W.T., et al.: Intelligent assessment of tractor frame performance based on machine learning. Mech. Des. **39**(03), 32–37 (2022)
2. Hou, Y.D., et al.: Research on the mechanism of enterprise digitalization affecting dual innovation performance. Res. Manag. **26**(3), 175–188 (2024)
3. Liu, D., et al.: Operation paradigm for remanufacturing shop-floor based on digital twin. Comput. Integr. Manuf. Syst. **25**(6), 1515–1527 (2019)
4. Leng, J., et al.: Digital twins-based remote semi-physical commissioning of flow-type smart manufacturing systems. J. Clean. Prod. **306**, 127278 (2021)
5. Li, H., et al.: The digital twin modeling method can be tested for automobile body-in-white welding production line. J. Zhongyuan Univ. Technol. **32**(01), 1–7 (2021)

6. Yan, D.X., et al.: Digital twin-driven rapid customized design of board-type furniture production line. J. Comput. Inf. Sci. Eng. **21**(3), 1–34 (2021)
7. Son, Y.H., et al.: Digital twin- based cyber-physical system for automotive body production lines. Int. J. Adv. Manuf. Technol. **115**(2), 1–20 (2021)
8. Liu, G., et al.: Artificial intelligence is the core engine of the development of new productivity in China. J. Hebei Univ. Econ. Bus. **45**(06), 61–71 (2024)
9. Wu, J.X.: Endogenous security issues and countermeasures for intelligent networked vehicles. J. Chongqing Univ. Posts Telecommun. (Nat. Sci. Ed.) **35**(03), 383–390 (2023)
10. Ma, X.H., Li, M.J., Qu, X., et al.: Research on gas turbine full life cycle digital twin based on system engineering. Therm. Power Eng. **37**(S1), 83–90+101 (2022)

Design and Realization of LoRa-Based Low-Power Smart Ranch Environment Monitoring System

Jiajie Tao and Yongfei Lu

Abstract *Objective* To solve the problems of high cost, high energy consumption, and insufficient transmission reliability of traditional ranch environmental monitoring and to provide powerful data support for real-time acquisition of ranch environmental parameters and ranch production management. *Methods* This system chooses the low-power and low-cost STM32WLE5CBU6 as the main controller and adopts the LoRa communication technology for long-distance data transmission combined with the 4G communication network, in order to realize the stable transmission of ranch environmental data, real-time monitoring, and remote management functions. On this basis, the system adds a low-power function to reduce the bandwidth required for data transmission by designing a data compression algorithm, which significantly reduces the overall energy consumption and makes it more suitable for deployment and long-term use in remote or difficult-to-frequently visit ranch environments. *Conclusion* The designed monitoring system is able to collect environmental parameters in real time and has the advantages of long transmission distance, strong anti-interference ability, and low cost and energy consumption, which is able to meet the needs of smart ranch environmental monitoring.

Keywords Smart ranch · LoRa · STM32 · Embedded environmental monitoring

1 Introduction

The livestock sector occupies a central position in China's agricultural system and has a significant impact on the national economy. As one of the world's largest livestock producers, China's total output value of the livestock industry in 2022 exceeded RMB 40,000 billion, accounting for about one-third of the total output value of

J. Tao (✉)
Shanghai Maritime University, Shanghai, China
e-mail: 869731574@qq.com

Y. Lu
East China University of Technology, Nanchang, China

K. Subramaniyam and M. Paprzycki (eds.), *Smart Cities and the Internet of Things*, Internet of Things, https://doi.org/10.1007/978-3-032-08290-9_14

agriculture, forestry, animal husbandry, and fisheries. Among them, the four major meat production categories of pigs, cattle, sheep, and poultry accounted for about 98 percent of the country's total livestock and poultry meat production. As of 2023, the country's pork, beef, sheep, and poultry meat production reached 96.41 million tonnes, an increase of 4.14 million tonnes, or 4.5%, over the previous year, of which 57.94 million tonnes of pork was produced [1].

Nonetheless, compared with China, developed countries in animal husbandry, like the Netherlands, have a higher proportion of livestock output value to total agricultural output value, close to 70 percent, showing that China's animal husbandry industry still has a lot of room for improvement in terms of output value growth [2]. With the progress of science and technology, many developed countries have made use of modern information technology such as Internet of Things, big data, and artificial intelligence to promote the development of animal husbandry in the direction of intelligence.

For the demand of smart ranch construction, a design scheme of LoRa-based smart ranch environment monitoring system is proposed. The scheme uses multi-sensor fusion technology to collect ranch environmental data and uploads these data to the cloud via LoRa network, ensuring the effective preservation of historical data. Ranch managers are able to remotely monitor environmental parameters in real time, which not only helps to adjust the ranch environment in a timely manner but also provides solid data and technical support for livestock health management and improved production efficiency. This LoRa-based monitoring system has become an effective tool for promoting the development of smart ranch informatization due to its low power consumption, long-distance transmission, flexible networking, and easy deployment.

2 General Structure of the Monitoring System

According to this study, the environmental monitoring system of the smart ranch system consists of LoRa terminal node, LoRa gateway, cloud platform, and personal computer (PC), etc. The LoRa terminal node collects the nitrogen, phosphorus, and potassium data of the soil; temperature and humidity data; the temperature and humidity data of the ranch environment; and the air CO_2 concentration data from the ranch with sensors and transmits the detected data to the IoT cloud platform through the LoRa gateway via the Message Queuing Telemetry Transport (MQTT) protocol, and the PC can obtain the data from the cloud platform to display and control the data. The PC can obtain the data from the cloud platform for display and control. When the environmental factors of the ranch change, the system sends commands through the cloud to control the output of related equipment, such as the opening of fans and pumps, so that the living environment of the ranch livestock can reach a balanced state. The cloud platform and PC provide the entrance to manually control the output devices, so that it is convenient for users to control the output devices

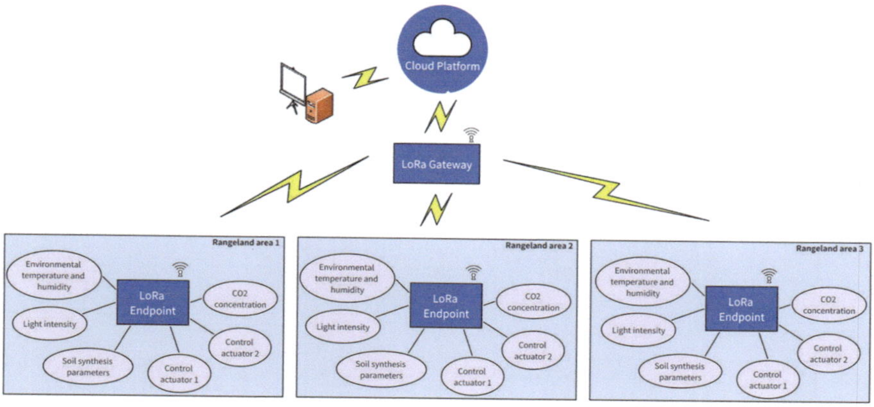

Fig. 1 System design block diagram

according to their own wishes, and the block diagram of the system design is shown in Fig. 1.

3 Hardware Design of the Monitoring System

3.1 Hardware Design of LoRa Endpoints and LoRa Gateway

The hardware part of LoRa endpoints and LoRa gateway consists of master control unit (MCU), data acquisition section, relay control section, power module, and network port module. The hardware framework is shown in Fig. 2. Actual installation diagram of LoRa endpoints and LoRa gateway is shown in Fig. 3.

The system adopts the STM32WLE5CBU6 high-performance low-power MCU, which supports multiple wireless modulation methods, including LoRa®, frequency-shift keying (FSK), and binary phase-shift keying (BPSK), and is suitable for long-distance communication in the 150–960 MHz band. The MCU is based on the Arm Cortex-M4 core, equipped with a floating-point unit (FPU) and integrates multiple peripheral interfaces (e.g., USART (Universal Synchronous/Asynchronous Receiver/Transmitter), LPUART (Low Power Universal Asynchronous Receiver Transmitter), I2C (Inter-Integrated Circuit), SPI (Serial Peripheral Interface), ADC (Analog-to-Digital Converter), and DAC (Digital-to-Analog Converter)), which improves the flexibility and scalability of the system.

For data acquisition section, VMS-3001-TR [3] soil integrated sensor is used to accurately measure soil temperature and humidity as well as nitrogen, phosphorus, and potassium content with high accuracy, strong stability, good salinity resistance, and anti-jamming ability and supports RS-485 bus and Modbus-remote terminal unit (RTU) protocols. In addition, VMS-3002-CO2 sensor and VMS-3002-GZ light

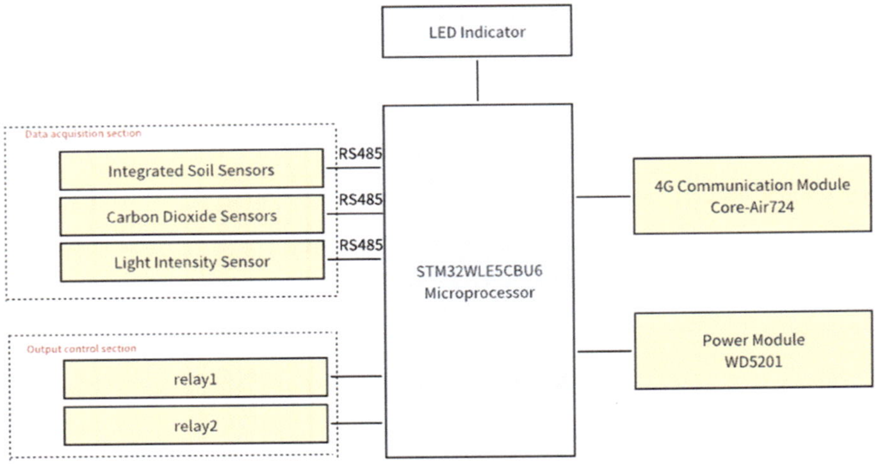

Fig. 2 LoRa endpoint and LoRa gateway hardware framework

Fig. 3 LoRa endpoint and LoRa gateway actual installation diagram

sensor are responsible for monitoring the air carbon dioxide concentration and light intensity respectively; both of them also support RS-485 bus and Modbus-RTU protocol to ensure the accuracy and reliability of data collection [4].

The system is also equipped with JQX-12F electromagnetic pump relay for precise control of irrigation, fertilizer, and spraying operations, which is widely used in farmland irrigation and other fields, featuring easy installation, small size, and low cost [5].

The power supply module adopts the WD5201 chip, which is capable of efficiently converting AC 220 V to DC 5 V, 3.3 V and 2.7 V without the need for additional power inductors or high-voltage capacitors, achieving compactness and cost optimization of the system.

For communication, the system is equipped with Core-Air724 4G communication module, which supports communication protocols such as MQTT and User Datagram Protocol (UDP), and is compatible with the 4G networks of Mobile, Unicom, and Telecom and has low-power mode and transient-voltage-suppression (TVS) protection to ensure the stability and safety of data transmission.

3.2 LoRa Communication Protocol

The communication between LoRa endpoints and LoRa gateway designs private communication protocols according to the product characteristics, LoRa gateway sends request frames to LoRa endpoints, and LoRa endpoints send response frames to LoRa gateway; the specific frame structure is shown in Tables 1 and 2.

Table 1 Request command structure

Variable	Length	Introduction
sync_head	2 bytes	Synchronized frame header, default value 0xAC 0x89
data_len	1 byte	Length of data field
crc	2 bytes	Checksum
frame_type	1 byte	Frame type, 0 x01 to read sensor data, 0x02 to control relay outputs
seq	1 byte	Frame sequence number
device_level	1 byte	Device level
dev_id	2 bytes	Device number
data	2 bytes	Control commands for relays

Table 2 Respond command structure

Variable	Length	Introduction
sync_head	2 bytes	Synchronized frame header, default value 0xAC 0x89
data_len	1 byte	Length of data field
crc	2 bytes	Checksum
frame_type	1 byte	Frame type, 0x01 to read sensor data
seq	1 byte	Frame sequence number
device_level	1 byte	Device level
dev_id	2 bytes	Device number
data	22 bytes	Collected data 4 bytes of soil temperature and humidity data 2 byte of soil ph data 6 bytes of soil NPK data 4 bytes of air temperature and humidity data 2 byte of light sensing data 2 byte of CO_2 sensing data

4 Software Design of Monitoring System

The software part of this system adopts KeiluVision 5 integrated development environment and is developed in C. Keil's interface is simple and clear and easy to operate, supports multiple programming languages, and is compatible with most Arm-based microcontrollers in the market.

4.1 Bootloader Computer Programming

The system has a built-in bootloader program, which is responsible for automatically detecting the server upgrade command during system startup and executing firmware download and update operations. During the upgrade process, the firmware integrity checking, breakpoint transmission, and rollback mechanism in case of abnormality are strictly implemented to ensure the safety and stability of the upgrade process. Flash partition is shown in Fig. 4.

The BOOT area is dedicated to storing the Bootloader, which is responsible for hardware initialization and loading applications; the UPGRADE_ADDR area is used to identify the system update status and decide whether to start the firmware upgrade process; the PARAM_ADDR_USER and PARAM_ADDR_FACTORY areas store the user-defined configuration information and the factory preset values, respectively, with the former providing a personalized experience and the latter serving as a backup when restoring the factory settings; the APP_ADDR area stores the running application code, which serves the core functions of the system; and the APP_ADDR_

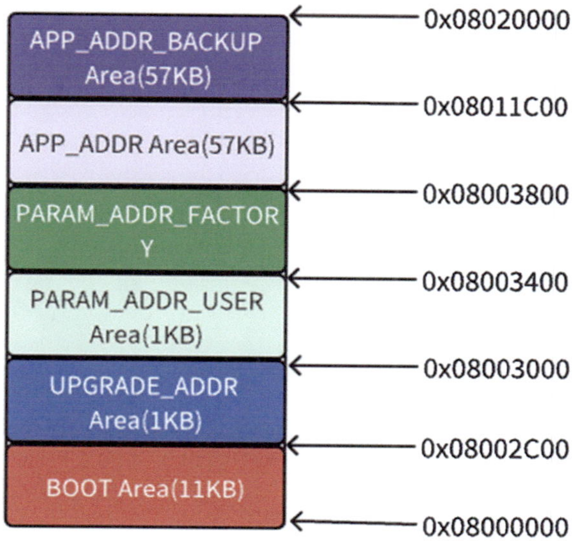

Fig. 4 Flash partition

BACKUP saves a stable version of the software, which allows a quick rollback in the event of an error in the new version and ensures the stability of the system.

4.2 LoRa Self-organizing Network Transmission

The LoRa self-organizing network of this system adopts a tree topology, where the lack of reliability of data transmission is a major obstacle in remote areas or environments with suboptimal signal coverage. The LoRa self-assembling network system with tree topology has significant advantages in such situations and helps to improve the reliability and continuity of data transmission. The LoRa self-organizing network with tree topology is shown in Fig. 5.

The system uses a cascading upload mechanism for long-distance data transmission. Each node maintains a list containing all the node IDs on the current path. The master control unit sends a data request to the superior node in a predefined format. The receiving node first verifies whether the data frame identifier matches its own, and if it does not, it checks whether it belongs to the nodes in the path list and then continues to forward the request after confirmation. Finally, the target node receives the request, collects the required data, encapsulates it into a message frame, and then returns it to the master control unit step by step along the original path. This approach ensures accurate data transmission while supporting long-distance communication. The long-distance data transmission method of cascading upload is shown in Fig. 6.

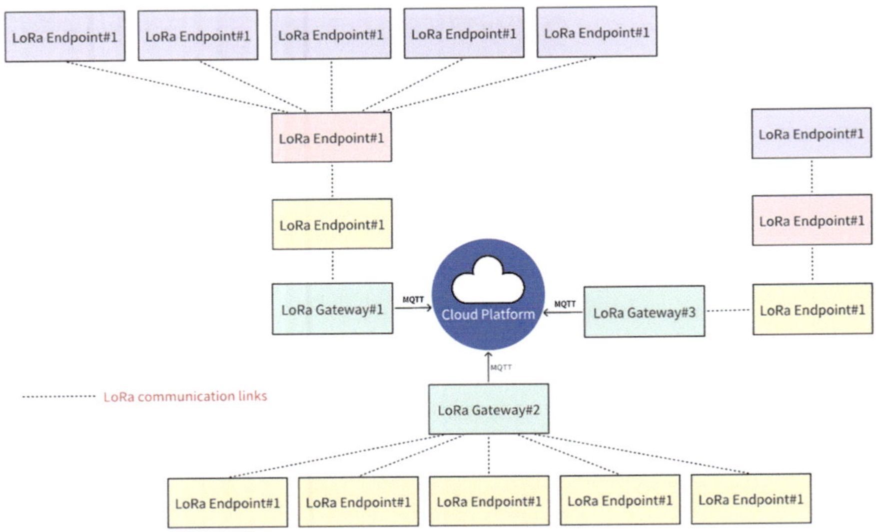

Fig. 5 Schematic diagram of LoRa self-organizing network

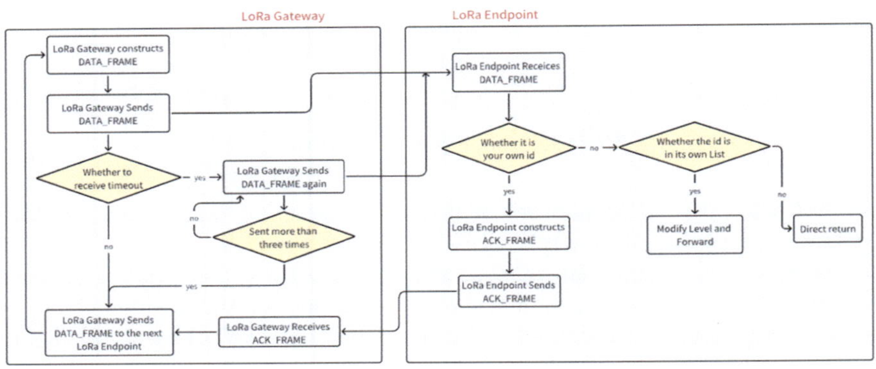

Fig. 6 Long-distance data transmission with cascading upload

4.3 *Low-Power Design*

4.3.1 Sleep and Wake Mechanisms

When the system is in idle state or there is no user operation, the system will be put into hibernation state; at this time, the power consumption is very low, and the quiescent current is only about 3.99 mA. When the LoRa terminal receives a wake-up frame from the LoRa gateway and matches the address of the node successfully, it will be woken up, which in turn wakes up the MCU in standby mode [6], and then,

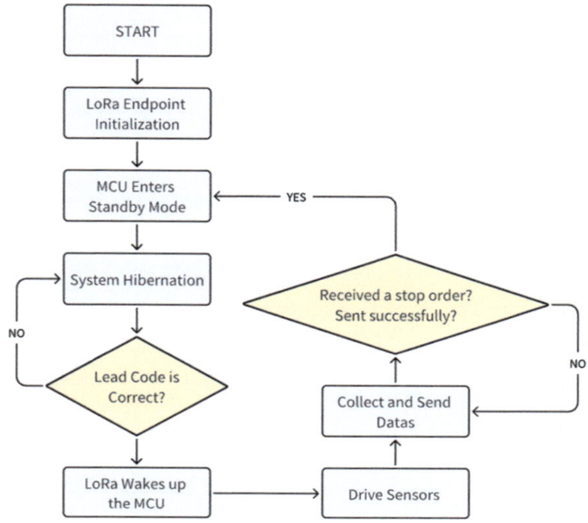

Fig. 7 The flowchart of the hibernation and wake-up mechanism

the MCU will drive the sensors to collect and send data. If the LoRa terminal receives a stop command from the gateway or has finished sending, then the MCU will enter standby mode again. If the leading code matching fails, then the system continues to remain in hibernation mode [7]. The flowchart of the hibernation and wake-up mechanism is shown in Fig. 7.

4.3.2 Lightweight Data Compression Algorithm Based on Run-Length Encoding (RLE)

The main energy consumption of LoRa devices is generated in the transceiver and data acquisition of wireless communication devices, which accounts for about 93.42% [8]. Therefore, reducing the length of LoRa transceiver data will significantly reduce the energy consumption of device operation [9].

Aiming at the characteristics of embedded systems with limited memory and processing power, this system designs a lightweight data compression algorithm based on RLE [10]. In the application scenario of smart ranch, the environmental data do not change much in a short period of time. Based on this feature, the algorithm first receives the first complete packet of environmental data at the receiving end as the baseline data. Subsequently, the collection end performs the heterodyne operation between each newly collected data and the previous benchmark data to find out the difference between them. Due to the small variation of the environmental data, the result containing a large number of consecutive zeros is often obtained after the heterodyne operation. For this reason, an improved RLE algorithm is used to further compress these consecutive 0's. The RLE algorithm is particularly suitable for processing data streams with a large number of repetitive characters, with high

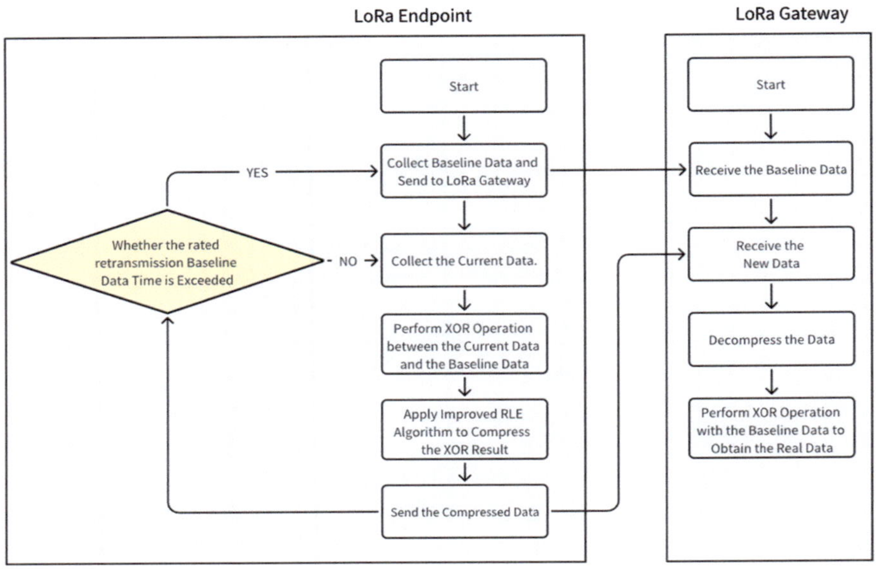

Fig. 8 The flowchart of the lightweight data compression algorithm based on RLE

computational efficiency, and is suitable for embedded systems with limited computational resources. In order to adapt to long-term environmental changes, the system updates the baseline packet every certain period of time to ensure the continuous effectiveness of the compression algorithm. This approach not only significantly reduces the amount of data to be transmitted but also maintains data integrity. The flowchart of the lightweight data compression algorithm based on RLE is shown in Fig. 8.

4.4 AliCloud Real-Time Monitoring

The main tasks of LoRa gateway include receiving data, uploading data, and displaying data in real time. Through LoRa technology, LoRa gateway receives the monitoring data sent by the collection nodes and then uploads the data to the AliCloud server through the 4G communication module [11]. The specific operation process of cloud platform communication is as follows:

- Product Creation: On the AliCloud IoT platform, you can select the appropriate category according to your needs, use direct-connected devices as the networking method, adopt 4G access, and use ICA data format (AlinkJSON) for authentication to ensure the safety and reliability of the product.

- Device Creation: After completing the product creation, the device adding interface pops up, and multiple devices can be added, each with unique device certificate information (triad information): ProductKey, DeviceName, DeviceSecret, and the MQTT protocol can be selected to communicate with Aliyun IoT. Establish a specific MQTT connection with Aliyun based on the triad information to activate the device.
- Attribute Creation: The device adds parameter attributes that need to be uploaded, such as temperature, humidity, light intensity, CO_2 concentration, soil temperature and humidity, soil PH, soil conductivity, and so on [12].

The cloud platform communication flow is shown in Fig. 9.

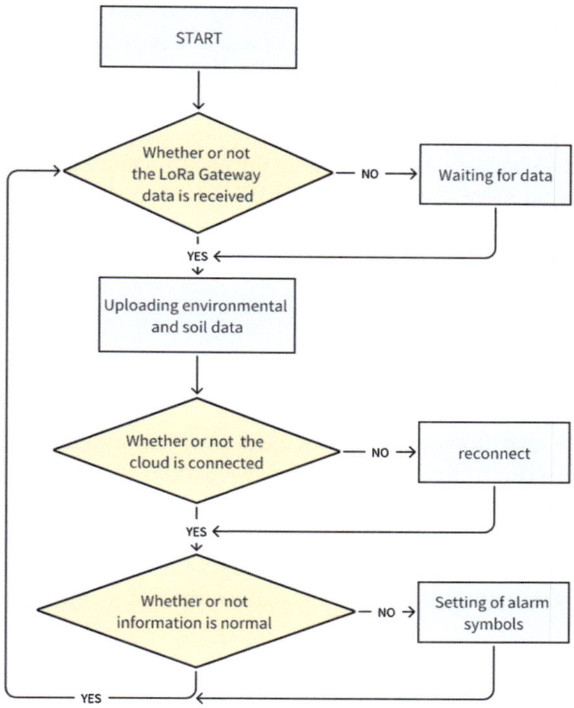

Fig. 9 The cloud platform communication flow

5 System Testing and Experimentation

5.1 System Wireless Communication Distance Test

The transmission distance of the wireless communication module is affected by the receiving sensitivity, which is related to the spreading factor, and the higher the spreading factor, the further the transmission distance of the wireless communication and the higher the power consumption. However, due to the low-power demand of the project, how to reduce the power consumption of the device as much as possible while ensuring the reliability of the data, the wireless communication distance test experiment was designed.

In order to ensure the stability and reliability of wireless communication, the LoRa modules of the gateway and nodes must be set with the same wireless channel band, bandwidth, and spreading factor. In this test, the air rate of the module was set to 0.3 Kbps, the transmit power was set to 22 dBm, the communication band was set to 500 kHz, and an endpoint alone was used for data communication with the gateway. After analyzing the experimental results, the maximum packet loss rate is no more than 1.5% within 100~1000 m for long-distance communication under sunny weather conditions, and the transmission distance and packet loss rate of this module can meet the project requirements.

5.2 Low-Power Test and Evaluation

In order to verify the effectiveness of the low-power design strategy of this system, a series of practical tests and evaluations were done on sleep and wake mechanism and lightweight data compression algorithm based on RLE, respectively. In one test cycle, the power consumption performance of the system in sleep and run modes is tested, as shown in Table 3, and the power consumption performance with and without the addition of the data compression algorithm is shown in Table 4.

Table 3 The power consumption performance of the system in sleep and run modes

Run-mode power consumption (W)	Sleep-mode power consumption (W)	Power consumption reduction rate
3.536	0.265	92.51%

Table 4 The power consumption performance with and without the addition of the data compression algorithm

System without data compression algorithm (W)	System with data compression algorithm (W)	Power consumption reduction rate
4.126	3.513	14.86%

Through comprehensive software design and effective data compression algorithms, the low-power design strategy of the system achieves the best results. The low-power design strategy of the system has achieved remarkable results. Experiments have shown that the power consumption of sleep mode is reduced by 92.51% compared to run mode, which means that sleep and wake mechanism can effectively reduce power consumption. Meanwhile, compared to the system without data compression algorithm, the system with data compression algorithm consumes 14.86% less power. This indicates that the low-power design scheme of this experiment has good feasibility and effectiveness in practical applications and can provide important support for the long-term stable operation of the smart ranch environment monitoring system.

5.3 *Environmental and Soil Data Acquisition Functional Test*

Accurate collection of environmental and soil data is essential for scientific research and plays an indispensable role in many fields such as agricultural production, environmental protection, and disaster prevention. The core objective of this test is to verify whether LoRa gateway can successfully receive the data transmitted by LoRa terminal devices.

As shown in Fig. 10, the LoRa gateway was able to accurately receive the data information collected by LoRa endpoint 1.

Fig. 10 Environmental and soil data acquisition functional test

```
START READ ALL SENSOR DATA :

LoRa Endpoint1:
          SOIL TEMPERATURE : 21.2
          SOIL HUMIDITY : 20.4
          SOIL PH : 6.88
          SOIL NITROGEN : 32mg/kg
          SOIL PHOSPHORUS  : 37mg/kg
          SOIL POTASSIUM : 48mg/kg
          AIR TEMPERATURE : 24.6
          AIR HUMIDITY : 35.9
          CO2(ppm) : 3100
          LIGHT INTENSITY : 15000Lx

success!
```

5.4 Environmental and Soil Data Acquisition Functional Test

After the deployment of the AliCloud IoT platform is completed, the smart ranch environment and soil parameters will be displayed in real time on the AliCloud IoT platform interface. After logging into the AliCloud IOT platform, the farmers can view the environmental parameters in real time, and the operator can set the threshold value according to the set threshold. According to the set threshold value, the operator can operate the JQX-12F equipment in the system debugging interface in time. The data display of AliCloud platform is shown in Fig. 11.

6 Conclusions

Aiming at the problems of high cost, high energy consumption, and unreliable data transmission in traditional ranch environmental monitoring, this study designs an environmental monitoring system based on STM32WLE5CBU6 low-power microcontroller, which combines LoRa and 4G communication technologies to achieve stable transmission, real-time monitoring, and remote management of ranch environmental parameters. In addition, the data transmission efficiency is optimized by introducing a data compression algorithm, which further reduces the energy consumption of the system and enhances its feasibility and economy for long-term use in remote areas. The experimental results show that the designed monitoring system is not only able to collect environmental parameters accurately in real time but also has the advantages of long-distance transmission, strong anti-interference, low cost, and low energy consumption, which fully meets the practical needs of environmental monitoring in smart farms.

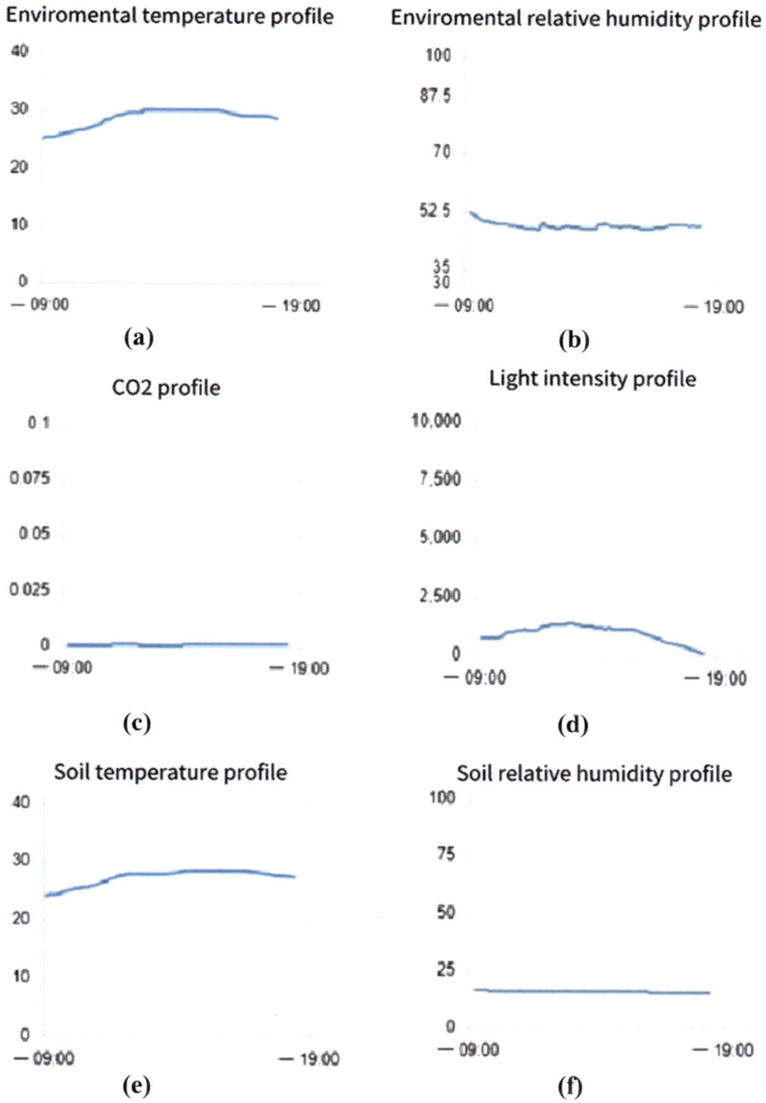

Fig. 11 AliCloud data showcase

References

1. Shu, Z.: The logic and progress of high quality development of animal husbandry enabling rural revitalization. Feed. Res. **47**(6), 185–189 (2024)
2. Thilagavathy, A., Ramesh, T., Selvi, S., et al.: Smart precision agriculture using LORA and WeIO interface. Carbohyd. Res. **8**(03), 3698–3701 (2019)

3. Sánchez Sutil, F., Cano-Ortega, A.: Smart public lighting control and measurement system using LoRa network. Electronics **9**(01), 124 (2020)
4. Ashifuddin M., Rehena, Z.: IoT based intelligent agriculture field monitoring system. In: Data Science & Engineering (Confluence), pp. 625–629 (2018)
5. Liya, M., Arjun, D.: A survey of LPWAN technology in agricultural field. In: 2020 Fourth International Conference on I-SMAC (IoT in Social, Mobile, Analytics and Cloud)(I-SMAC), pp. 313–317 (2020)
6. Babić, D., Jovović, I., Popović, T., Kovač, N., Čakić, S.: An internet of things system for environmental monitoring based on ESP32 and Blynk. In: 2022 26th International Conference on Information Technology (IT), pp. 1–5. Zabljak, Montenegro (2022)
7. Kasmeera, K.S., James Shine, P., Sreekumar, K.: Efficient compression of secured images using subservient data and huffman coding. Procedia Technol. **25**, 60–67 (2016)
8. Ahmed, T., Bristy, S.K., Fahad, A.A., Mia, M.S.: Smart decision maker and monitoring system for modern agriculture based on internet of things. In: Communication, electrical, and smart systems (iCACCESS), pp. 1–6 (2024)
9. Jin, M.-J., Qu, Z.-W.: Efficient neighbor collaboration fault detection in WSN. J. China Univ. Posts Telecommun. **18**(SUPPL.1), 118–121 (2011)
10. Yu, Z., Xiao, B., Zhou, S.: Achieving optimal data storage position in wireless sensor networks. Comput. Commun. **33**(1), 92–102 (2010)
11. Zhao, J.-D., Wang, Q., Wan, Y.-D.: A channel adaptive algorithm for industrial wireless sensor network with mesh structure. Dongbei Daxue Xuebao/J. Northeast. Univ. **31**(SUPPL. 2), 214–219 (2010)
12. Shoaib, S., Chen, X., Llewellyn, I.: Increasing the effective range of smart meter home area network. IEEE Antennas Wirel. Propag. Lett. **16**, 2898–2901 (2017)

A New Smart Home System Based on IoT from 5G

Zhaoye Li, Jiali Quan, and Mengying Liu

Abstract In this study, a new smart home system was designed and discussed. In order to solve fire warning and avoiding disaster damage, it was assembled by flame sensor, temperature and humidity sensor, smoke sensor, and Hall sensor. The signals collected by the sensors can be transmitted to the phone using 5G technology, which can prompt the temperature, humidity, and smoke status of the living environment. It is very useful and healthy for modern people.

Keywords Internet · 5G · 6G · Internet of Things technology · Smart home

1 Introduction

In the history of human development, housing has always been one of the goals that people pursue. Owning a home is considered a status symbol. In the Stone Age, people lived in caves and did not forget to use stones and flowers to decorate their homes, which will bring happiness for them. In feudal times, employers or emperors would hire some servants to provide various housing services, such as safety opening and closing windows for their residences.

With the improvement of smelting and firing technology, brick and cement have become the main building materials. And people's houses are becoming stronger and stronger. However, it is still impossible to give early warnings and notices of disasters such as fire. Even today, with the development of science and technology, thousands of people die in fires every year. In this context, alarm-type houses that can predict disasters in advance have become one of the pursuing goals.

These alarm-type houses rely on fast response Internet and communication technology. In 1984, the United States creatively designed the first smart housing project [1]. Then, Canada, Australia, and other countries had taken action to put forward diversified smart housing programs. At this time, smart housing is popular around the world. However, China's Internet technology started late a little. According to

Z. Li (✉) · J. Quan · M. Liu
Tianjin Vocational and Technical Normal University, Tianjin, China
e-mail: xinsuifandong@tute.edu.cn

K. Subramaniyam and M. Paprzycki (eds.), *Smart Cities and the Internet of Things*, Internet of Things, https://doi.org/10.1007/978-3-032-08290-9_15

different roles till now, the development of China's Internet can be divided into six generations [2].

Its first generation, also known as the analog voice era, was content to make simple calls using mobile phones (such as Motorola and Ericsson). In the second generation, the text age, people can not only make phone calls but also send text messages, which is a remarkable leap forward. In 1990, a phone can start surfing the Internet, but it can only read text messages. Its download speed is only about 20 K/s, which is not satisfied with large-scale data transmission.

In the third generation, mobile communication has entered the era of pictures. At this time, it is not only to make phone calls and send text messages but also to realize the combination of wireless communication and multimedia communication. At the same time, the competition of foreign mobile phone manufacturers is fierce. Certainly, the elimination rate is high. The third generation did not last very long before entering the 4G era.

The fourth generation is also known as the video age. After 2010, the mobile Internet can transmit high-quality videos and images. Its image transmission quality and clarity can be comparable to television technology products. This is also the basic condition of the Internet of Things (IoT) technology [3]. It contains three layers of structure, namely perception layer, network layer, and application layer [4].

Since 2019, China has entered the era of the Internet of Everything, meaning the fifth generation (5G) of the Internet. 5G is not only a single wireless access technology but a general term for a variety of new wireless access technologies and the integration of existing evolving technologies. It is the integration of communication technology and the Internet in the true sense. Mobile phones are connected to everything, such as car autonomous driving at this time. The 5G era is also known as the Industrial Internet era. The 6G network latency that is now being developed may range from milliseconds to microseconds, which greatly improves the level of industrial interconnection.

The Internet of Things is the technical foundation of smart homes [5]. It means a network based on information carriers such as the Internet and traditional telecommunications networks. In this condition, all ordinary physical objects that can be independently addressed can be interconnected. It has three important characteristics: common object device, autonomous terminal interconnection, and universal service intelligence.

In 2009, Internet of Things technology was firstly raised by China Mobile Communications Corporation. Till now, the number of IoT connections in China is expected to exceed 3 billion. There are many different successful applications about IoT, such as smart homes, smart healthcare, smart logistics, and so on [6].

Although the development of the Internet of Things has gradually become a scale in recent years, countries have invested huge human, material, and financial resources to carry out research and development. However, there are still many problems to overcome in terms of technology, management, cost, policy, security, etc.

In this study, a new kind of smart home was designed according to IoT. Its key technology is 5G technology for high-speed data transmission. Many different signals can be caught by various sensors. After single-chip processing, all the detection results will be shown in mobile phone interface.

2 Principle and Methods

2.1 The Flowchart of Operating Principle

The smart home system is designed according to the principle of IoT based on 5G technology [7]. It is carried with STM32 single-chip system as central control unit. Its construction is shown in Fig. 1. Every sensor is very much sensitive. For example, temperature and humidity sensor will give an alarm once the temperature and humidity of the room exceed the threshold. Flame sensor can monitor the room for open flames. If the smoke in the room exceeds the limit, the smoke alarm will work immediately.

Hall sensor is a kind of magnetic field sensor based on Hall effect. Its main function is to identify the phase position information of the motor winding and convert it into an electrical signal. The driver obtains the position information of the rotor by reading the level signal of the output end of the Hall element. The logic switch completes the correct reversal according to the position information of the rotor of the motor.

In the whole smart system, Hall sensor is connected with motor and Universal Serial Bus (USB) light. According to the intensity of indoor light and the requirements

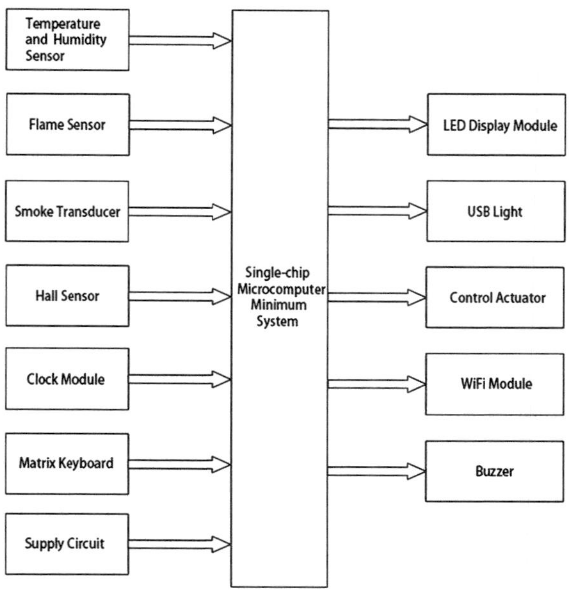

Fig. 1 System chart for smart home

for lighting intensity, it is satisfied with different requirements for family members with various age-groups.

At the same time, it is connected with curtain. Relying on Hall sensor, motor can control curtain according to weather or light intensity.

There is a WiFi module, which is an information transmission module based on 5G technology. WiFi technology is often used in smart furniture because of the fast data transmission rate. It has a frequency band of 2.4 GHz or 5 GHz. There are three characteristics, namely short transmission distance, fast transmission rate, and wide coverage. This integration enhances the practicality and usability of the samrt home. According to the literature [8], the WiFi transmission rate formula is shown formula 1, where R represents WiFi signal transmission rate, B represents channel bandwidth, and SNR signifies signal-to-noise ratio. It is shown that the higher channel bandwidth and SNR will bring faster signal transmission rate. In this smart system, B was chosen as 5G Hz due to its high transmission speed and low interference.

$$R = B \log_2^{(1+\mathrm{SNR})} \tag{1}$$

In order to improve SNR, wider information channel is a good choice. Besides, the access point (AP) is as close to the user terminal as possible. Certainly, avoiding wall connections is also a fair option.

Bluetooth technology is also a kind of Internet of Things link technology [9–11]. It is an open short-range wireless communication standard. Through the wireless connection, it can achieve direct, flexible, secure, low-cost, and low-power network interconnection between fixed devices and mobile device communication environments in a small range. It can also realize voice and data communication between various digital devices.

Bluetooth technology works in the universal 2.4 GHz band. In theory, the 2.4 GHz operating technology can cover devices within 30 m. But in reality, its coverage can reach only about 10 m. Within the effective range, it can achieve eight devices connected to each other at the same time. Within a distance of 10 m, the receiving device can dynamically adjust its power. When the missions are reduced or stopped, the Bluetooth device can enter the low-power mode. In low power Bluetooth mode, its transmission distance is calculated as shown in formula 2, where P_t represents transmission frequency, G_t represents transmitting antenna gain, G_r represents receiving antenna revenue, λ represents wave length, and L represents path loss.

$$d = \sqrt{\frac{P_t G_t G_r \lambda^2}{(4\pi)^2 L}} \tag{2}$$

In smart home system, Bluetooth Low Energy works normally. Through frequency hopping technology, the signal is divided into different frequency bands, which ensures the stability of the link and reduces interference.

Figure 2 shows the flowchart of operating principle for the smart home, which explains the relationship between sensors and transmission technique. Through radio frequency identification technology (RFID), lock status is monitored seriously. If it is not closed, it will enter closing time with the help of motor controlled by Hall sensor. Then, the system will monitor the house for anyone relying on infrared detection technique. The result will be shown in mobile phone interface. The presence or absence of fire will be monitored by other sensors containing temperature and humidity sensor, smoke sensor, and flame sensor.

Figure 3 shows the flowchart of the smart home system for data processing through the cloud platform. As a reliable intelligent housing system, the system will monitor various sensors' data every 500 ms. All data will be sent to cloud platform. After data comparison and result analysis, the final decision on whether to alarm is made.

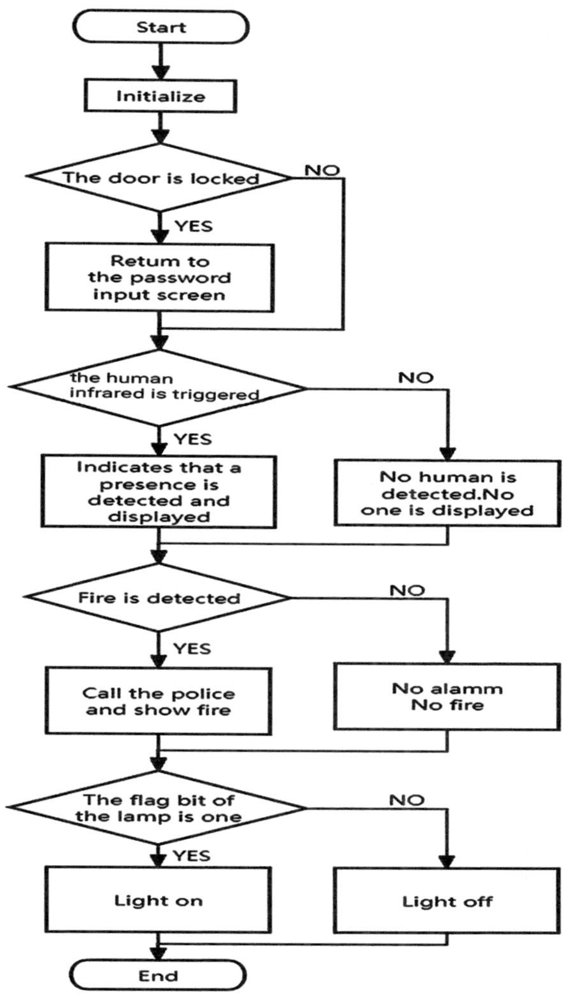

Fig. 2 Flowchart of operating principle for the smart home

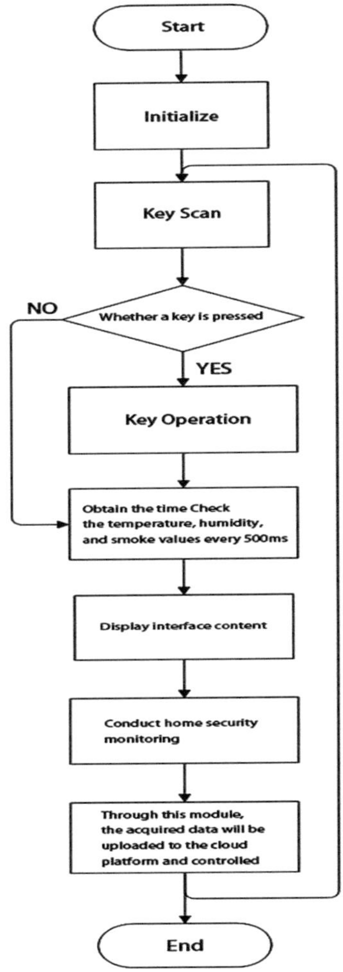

Fig. 3 Flowchart of the smart home system for data processing through the cloud platform

2.2 *Principle*

In this system, Huawei Cloud serves as the data storage and computing center. Its sensing layer is composed of temperature and humidity sensor, flame sensor, smoke sensor, and Hall sensor, which are responsible for collecting data. The network layer is composed of Bluetooth low power transmission module in STM32 and WiFi module.

Data transmission protocols based on China Unicom 5G adopt Internet Protocol version 6 (IPv6) as a transport layer, which can provide more IP address space for support large-scale device linking. Hypertext transfer protocol (HTTP) and real-time transport protocol (RTP) can be responsible for defining the data formats and communication rules exchanged between communication parties. Its fast data transfer speed provides support for the development of the Internet of Things.

Huawei Cloud is very helpful for the smart home system. Its high safety can ensure effective access by legitimate personnel through technologies such as face recognition. It can also help users access valid object creation. At the same time, log access records can help owners view the security of the house and intrusions.

3 Results and Discussion

In this system, experiments are arranged for testing sensor effectiveness. The data are shown in Table 1 and in Fig. 4 as an example.

In fact, different kinds of materials display different smoke testing value and smoke warning value. In the environment of our laboratory, when the lighter is lit for a while, its smoke warning value can reach 42 parts per million (ppm). Most of the burning in lighters is liquefied gas, which produces carbon dioxide when the air is sufficient. It can cause an alarm from the smoke sensor. In reality, the smoke sensor can also detect presence of carbon monoxide, whose alarm threshold maybe reaches 600 ppm. In order to prevent danger, burn test and smoke test are arranged by using ready-made cigarette butts. When it is burning, its threshold value can reach 192 ppm. Without burning, the air smoke value is tested as 12 ppm.

In the temperature test shown in Table 1, temperature is different from indoor actual temperature. That is because of measurement error of temperature sensor.

From Fig. 4, it can be seen that the value in the system display is the same as one in 5G cell. This means effectiveness of reliable data transmission over short distances. In the experiment, the Bluetooth on the STM32 is connected to the Bluetooth of the phone. At the same time, data can also be transmitted by WiFi. Double insurance ensures the validity of data transmission.

The fast 5G transmission speed is connected to Huawei Cloud. All the experiment data can be stored in Huawei Cloud. When the buzzer sounds, the peak value maybe indicates alarm category by consulting log in Huawei Cloud.

In order to ensure the security of the smart home, the access control system uses radio frequency identification technology. By brushing the access card, it can let the family members enter. This can play a certain security protection.

Table 1 Data about loading experiments

Indoor temperature (°C)	Test temperature (°C)	Type of smoke sensitivity	Smoke value before test/set smoke threshold (ppm)
29	29.1	Cigarette butts	12/192

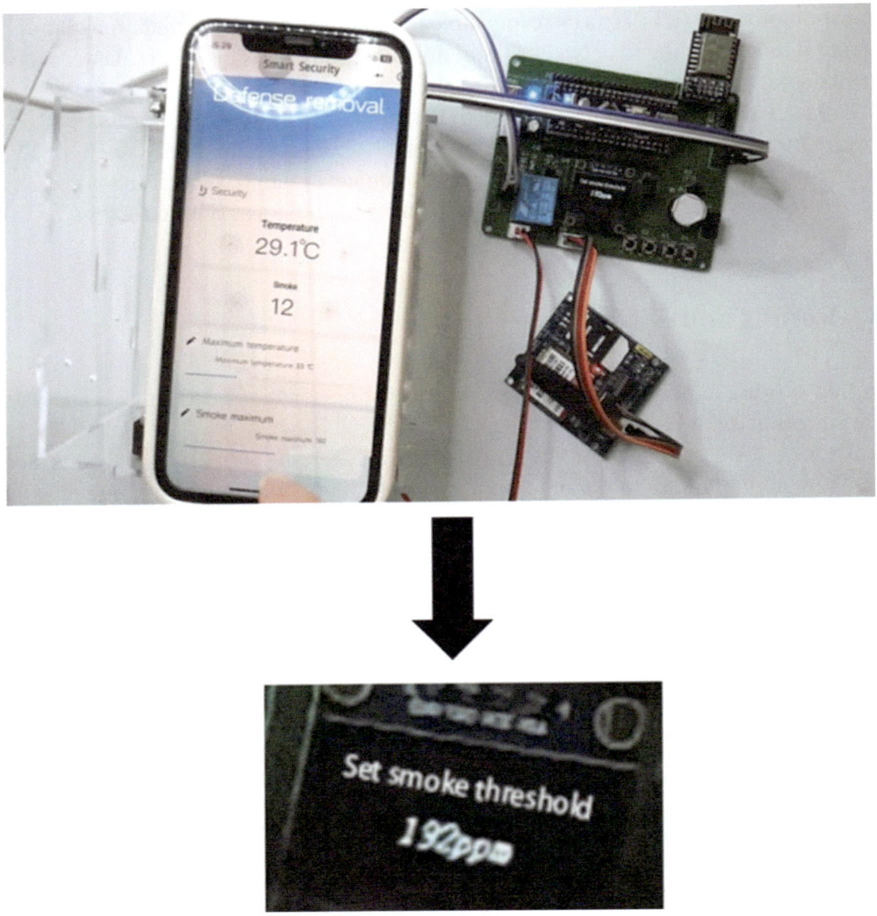

Fig. 4 The test result displayed by China Unicom 5G technology

4 Conclusions

In this study, a new smart home system was designed and discussed. All sensors can feel the changes in your environment. These changes can be sent to Huawei Cloud and displayed on 5G mobile phones. Through the Internet of Things connection based on 5G, the home can play a timely alarm role in the fire. Huawei Cloud records can be checked by the owner.

Acknowledgements This work was supported by Tianjin municipal education committee higher education development fund (JWK1619) and Tianjin municipal college student entrepreneurship training program project funding (202410066153).

References

1. Chen, J.W., Huang, Z.M., Cai, Z.Z.: Device side smart home service recommendation framework based on deep learning. Appl. Res. Comput. **41**, 533–539 (2024)
2. Tian, M.J.: China's internet development and evolution trend and application status. Theory Chin. Commer. **27**, 22–24 (2018)
3. Guan, G.L., Zhi, T., Tao, Z.P., Cao, Y.: A federated learning privacy protection method for multi-key monomorphic encryption in the Internet of Things. J. Inf. Secur. Res. **10**, 958–966 (2024)
4. Chen, X.D.: Analysis of application forms of Internet of Things and smart home control technology. Inf. Comput. **13**, 124–126 (2020)
5. Zhu, M.W., Zhu, B.G., Liu, G.X., Nan, R.T.: A teaching system for smart home based on Internet of Things. Inf. Technol. **5**, 21–25 (2011)
6. Li, H.Y.: Quality management and performance evaluation of smart home system based on Internet of Things technology. Value Eng. **4**, 34–36 (2024)
7. Liu, D., Qin, Q.S.: Research on the combination of 5G technology, fog computing and smart home. China Plant Eng. **23**, 39–40 (2021)
8. Chen, L.: The application of 5G communication technology in smart home design. Telecom Power Technol. **41**, 25–27 (2024)
9. Xiao, J.H., Wu, J.T.: Bluetooth technology and smart home system design. Radio Eng. **34**, 52–55 (2004)
10. Wang, L., Huang, G.M., Zhao, L.: Bluetooth key technology and smart home. Appl. Single Chip Microcomput. Embed. Syst. **9**, 7–9 (2003)
11. Peng, C.H., Zhou, Y., Chen, G.B.: IoT smart home control system based on Bluetooth Mesh and NB-IoT. Electron. World **24**, 154–155 (2020)

Design and Implementation of Intelligent Kitchen Environment Monitoring System

Yuting Jiang

Abstract Kitchen is a gathering area of grease and gas; the traditional kitchen environment is poor. People judge the kitchen environment through experience. In order to improve the kitchen environment, a smart kitchen environment monitoring system is designed based on the background of the Internet of Things (IoT). The system is based on the STM32 as the core, including the two parts: hardware side and the app side. The hardware side can monitor in real time the temperature and humidity, flooding, smoke, combustible gases, flames, and other data. The app side can monitor the kitchen environment, view the collected data of the kitchen environment, and monitor and control the kitchen environment manually or automatically through the threshold control, so as to realize the intelligentization of the kitchen environment monitoring. The wireless communication of the system is realized by using Zigbee technology, realizing multinode division of labor processing, and the interface is clear and easy to use.

Keywords STM32 · App · Zigbee · IoT · Kitchen

1 Introduction

The kitchen is a relatively dangerous place in family life, often experiencing safety issues such as cooking fires, gas leaks, and excessive smoke [1]. To address these problems, an intelligent kitchen environment monitoring system has emerged [2]. The main focus of this research is to collect data through a sensing layer, which then automatically executes operational controls after threshold judgments. Additionally, in manual mode, users can monitor and adjust various environmental parameters in the kitchen in real time through an app. The system is also designed with multiple nodes in consideration of the characteristics of potential water leaks and gas leaks in different areas of the kitchen.

Y. Jiang (✉)
Department of Network and Internet of Things Engineering, Dalian Neusoft University of Information, Dalian, China
e-mail: Jiangyuting@neusoft.edu.cn

K. Subramaniyam and M. Paprzycki (eds.), *Smart Cities and the Internet of Things*, Internet of Things, https://doi.org/10.1007/978-3-032-08290-9_16

2 Related Technologies

2.1 STM32

The STM32 is a powerful 32-bit ARM microcontroller that features high performance and low power consumption, as well as a rich set of peripheral interfaces [3]. It is widely used in the field of embedded system development and offers a wealth of development tools and ecosystem support. It is particularly suitable for the implementation of smart kitchen environmental monitoring systems.

2.2 Zigbee

Due to the complexities of wiring, high costs, and maintenance difficulties associated with traditional wired systems, new IoT systems tend to use wireless connections [4]. The Zigbee technology used in this system features short range and low power consumption, making it suitable for small sensor systems. This system uses the CC2530 to call sensor perception layer data, which is measured and sent to the coordinator and then forwarded to the core microcontroller unit (MCU) for processing.

2.3 Android Studio

Android Studio is an open-source mobile operating system developed by Google, based on Linux [5]. With its excellent performance and wide range of application scenarios, it has become the most widely used mobile operating system in the world. This system is based on Android technology and has successfully developed an intelligent kitchen environment monitoring system [6]. It aims to provide users with a convenient and real-time monitoring experience. The app interface is designed to be simple and clear, with practical functions that offer users significant practical value.

3 System Design

3.1 Requirements Analysis

- The Android app can view data such as temperature and humidity, gas concentration, flames, and water immersion for each node and can perform visual analysis and processing of the data.

- It enables remote control of various devices in the kitchen, such as fans, water pumps, buzzers, and other actuators.
- The system needs to be able to monitor changes in temperature and humidity in the kitchen in real time and to issue alarms and allow for remote control in case of abnormal environmental conditions, ensuring a comfortable and safe kitchen environment.
- The system needs to monitor smoke and fire conditions in the kitchen and be able to issue timely alarms and take appropriate firefighting measures to ensure kitchen safety.
- It should provide user management functions, such as user login and registration.

3.2 System Design

The system consists of three parts: the Android client, the hardware component, and the cloud platform. The cloud platform allows for real-time data display and monitoring of the actuator's working status. The hardware component includes an STM32 development board, an ESP8266 Wi-Fi module, a CC2530 Zigbee module, a DHT11 temperature and humidity sensor, a smoke sensor, a flame sensor, a CO sensor, a water immersion sensor, as well as fan modules, pump modules, and buzzers. The STM32 main control board connects to the network via the Wi-Fi module and communicates with the OneNET cloud platform using the MQTT protocol, enabling data transfer between the hardware component and the Android client. The main node and each subnode communicate and exchange messages via Zigbee, as shown in Fig. 1.

(1) **Hardware Design**

(a) *Sensor and Controller Design*
The sensor selection for the intelligent kitchen environment monitoring system is mainly based on the different monitoring parameters and monitoring functions. The selected sensors include DHT11 temperature and humidity sensor, flame sensor, MQ-2 smoke sensor, CO sensor, and water level sensor. The controller is designed with relays, buzzers, water pumps, and fans.

(b) *Transmission Module Design*
The ESP8266 is chosen to transfer data to the cloud [7]. Because of the low power consumption and high flexibility characteristics of Zigbee, CC2530 is used to realize wireless transmission networking.

The system is divided into two modes: manual and automatic. The collected data is evaluated and transmitted to the app after uploading to the cloud platform. Users can choose between the two modes. The flowchart of the main modules of the hardware is shown in Fig. 2.

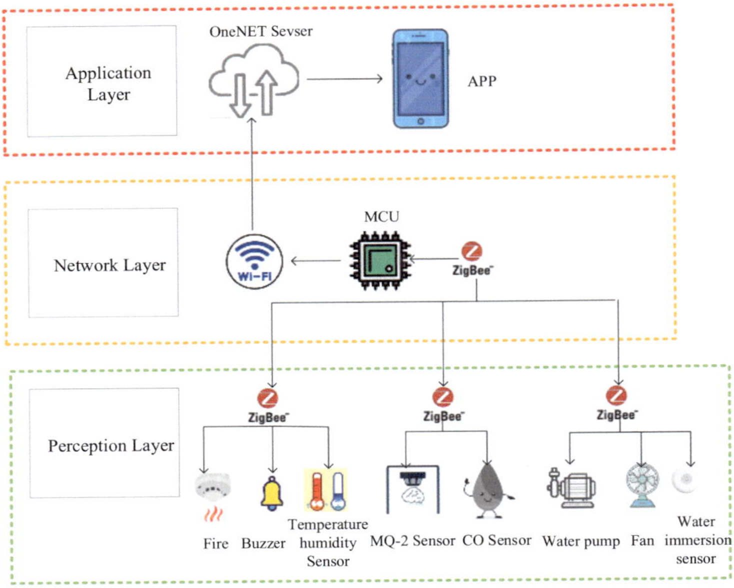

Fig. 1 System architecture

Fig. 2 Hardware design flowchart

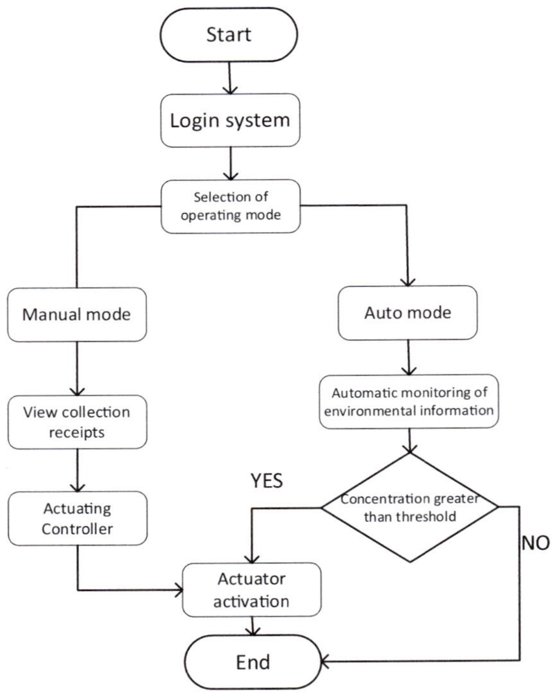

Table 1 Software module structure

Module		Functionality
UI		Provides UI presentation interface
User-function layer	Intelligent control mode selection	Choose manual or auto module
	Manual module	Remote control of kitchen actuators
	Auto module	System works automatically according to the collected data
	Information viewing	Provides information on kitchen temperature and humidity, smoke, flames, combustible gases, etc
	Alarm tips	Exceeds design threshold alarm
	Historical data	View all historical data
Network communication	Zigbee, MQTT, Wi-Fi	Provides wireless communication function

(2) **Software Design**
Software part of the design UI interface display, divided into two modes, need to include two control modes, information view and historical data view pinning function, APP Brother layered module function table shown in Table 1.

4 Implementation

4.1 Hardware Implementation

The system is based on the STM32 core gateway, paired with Zigbee to drive the DHT11 sensor, CO sensor, MQ-2 sensor, flame sensor, and water immersion sensor to collect data. After processing the data, it automatically activates the actuator to adjust the environment. The data is transmitted to the cloud via ESP8266 for viewing on an app. The hardware module is shown in Fig. 3.

4.2 Android Studio Implementation

(1) **Register and Login**

When the app is opened, the first thing that appears is the app login interface, where users can log in with their preregistered username and password. The register and login interface is shown in Fig. 4.

(2) **Information Display and Threshold Modification**

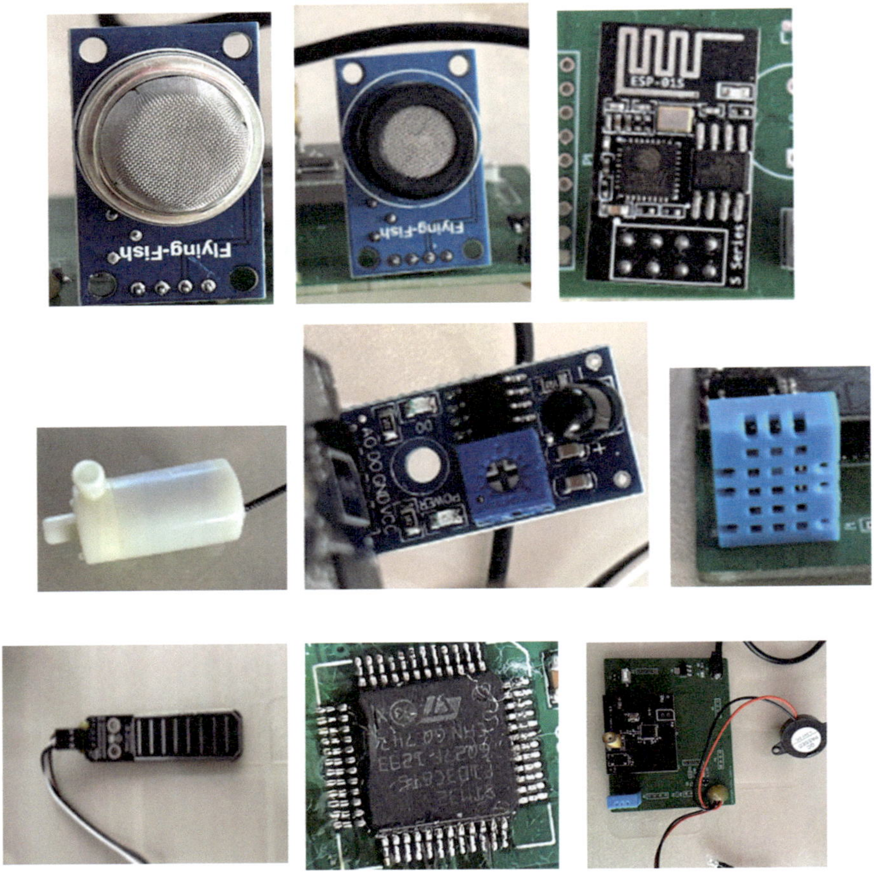

Fig. 3 Hardware module

Users can view the real-time display of data and real-time control of the actuator through the monitoring and control page. Real-time data monitoring allows users to view the current environmental temperature and humidity, combustible gas concentration, smoke, water level, flame, and other real-time data. Users can also set the thresholds for modification of the system to respond to the real-time data monitoring as shown in Fig. 5.

(3) **Historical Data Query**

Users can query and analyze the historical information through the information query page of the app interface and can clear historical information records. Historical information shows various real-time data tested at a certain period of time. The historical data query interface is shown in Fig. 6.

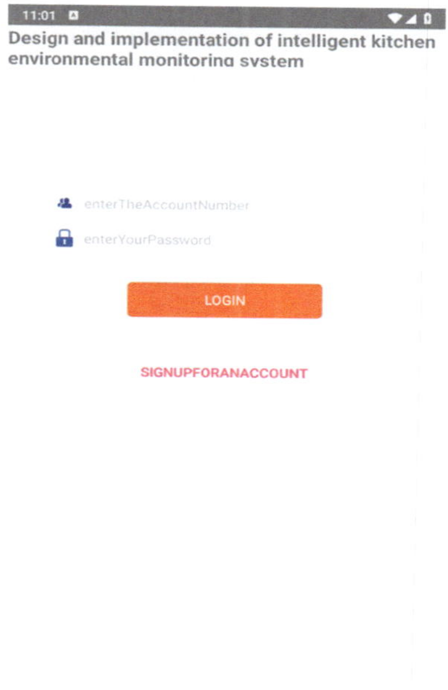

Fig. 4 Register and login interface

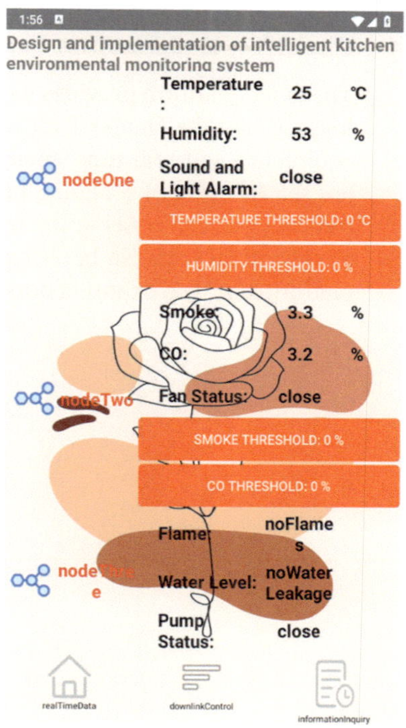

Fig. 5 Information display and threshold modification

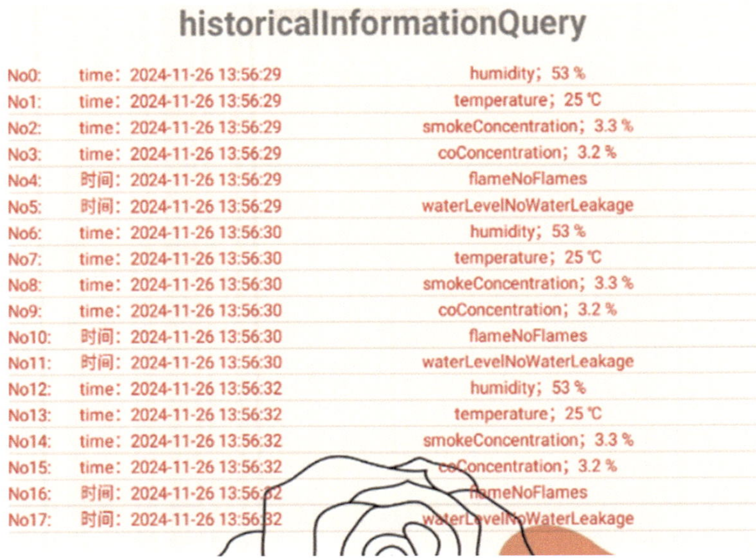

Fig. 6 Historical data query

(4) **OneNET Cloud Platform Functional Modules**

The cloud platform provides the function of viewing data after logging in. After logging into the OneNET cloud platform, users can view the real-time online equipment and real-time environmental monitoring data display, and it also has the function of retaining historical data. In addition, the user can click on the data list, can observe the sensor before the monitoring of all the data, and the historical data can be visualized as a line graph, easy to watch. The cloud platform can also export historical data. The specific display is shown in Fig. 7.

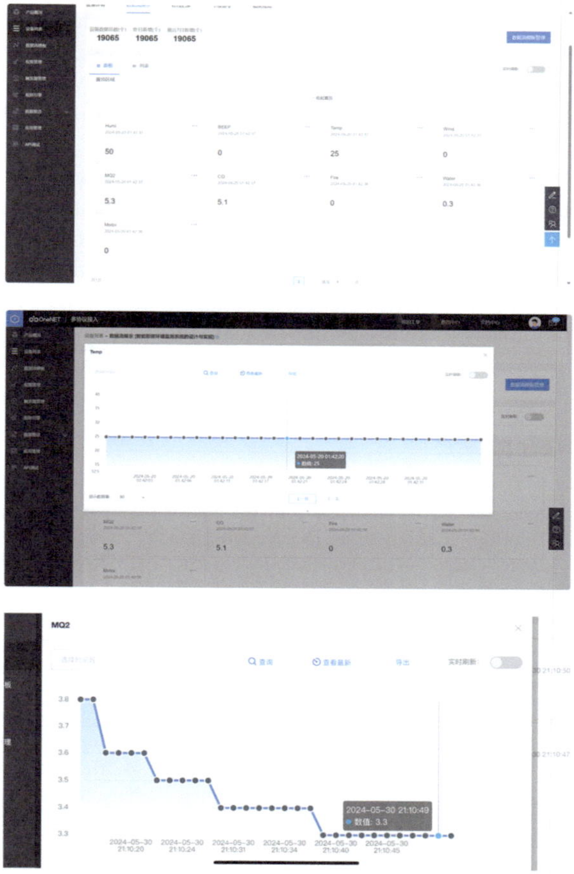

Fig. 7 Cloud platform historical data

5 Conclusions

The design of the intelligent kitchen environment monitoring system integrates technologies such as wireless communication, sensors, and data collection and analysis. This system is capable of monitoring combustible gases, temperature, humidity, and other factors in the kitchen environment. By uploading data and processing it, the system provides real-time and accurate monitoring results. Additionally, users can remotely control the actuators in the system through devices like smartphones based on the current kitchen environment, greatly enhancing kitchen safety. Testing has shown that the system has advantages such as safety, stability, and reliability, while also being cost-effective and easy to operate, fully meeting users' needs and ensuring kitchen safety.

Although the system has completed initial development and testing, there are still some issues in practical applications. First, aspects such as hardware selection can be further improved, and the types of monitoring data are relatively limited. Second, the system's handling of extreme situations needs optimization, such as power outages and poor network conditions; currently, the system cannot effectively respond to these scenarios. Last, the cost and lifespan of the system must be considered, aiming to enhance sustainability while reducing costs.

In summary, the intelligent kitchen environment monitoring system still has significant room for improvement and more application scenarios. The project development phase faces some challenges, but in order to ensure safety in home environments and create an intelligent living space, these challenges must be addressed. This will enable the intelligent kitchen environment monitoring system to advance further in the rapidly evolving technological landscape.

References

1. Sandeep, M., Nandini, C.: Implementation of IoT based smart cooking environment (2019)
2. Jian-Mei, S., Li-Juan, W., Xi, Y., Jun-Shan, Y.: Design and implementation of intelligent kitchen system based on Internet of Things. In: 2018 International Conference on Smart Grid and Electrical Automation (ICSGEA), pp. 253–256. Changsha, China (2018)
3. Wu, S., Abdulghani, A.M., Ansari, S., Imran, M.A., Abbasi, Q.H.: IoT enabled smart lighting system using STM32 microcontroller with high performance ARM® Cortex®-M3 core. In: 2020 International Conference on UK-China Emerging Technologies (UCET), pp. 1–4. Glasgow, UK (2020)
4. Pawaskar, M., Aneesh, S., Sharma, D., Sawale, P., Sharma, R., Sharma, A.: IoT enabled Smart Dustbin using Zigbee Network. In: 2021 International Conference on Advances in Computing, Communication, and Control (ICAC3) pp. 1–4. Mumbai, India (2021)
5. Ch, J.: An android application for smart garbage monitoring system using Internet of Things (IoT). In: 2023 7th International Conference on Computing Methodologies and Communication (ICCMC), pp. 1242–1248. Erode, India (2023)
6. Sheela, G.K., Jilna, K.: An intelligent energy management system using IOT. Mater. Today Proc. **24**(Pt 3), 1903–1908 (2020)
7. Rosli RS, Habaebi MH, Islam MR (2019) On the analysis of received signal strength indicator from ESP8266. Bull. Electr. Eng. Inform. **8**(3), 933–940

Design of Intelligent Operation and Maintenance System of Ticket Sale and Inspection Equipment of Urban Rail Transit Station Based on PHM Technology

Haitao Deng, Xiali Cao, Hao Li, and Yanlan Meng

Abstract With the rapid development of the economy and the continuous improvement of people's living standards, urban traffic congestion, low efficiency of mass travel, and other problems are becoming more and more obvious. Urban rail transit is an efficient, safe, and green means of transportation, which is the best way to solve urban traffic congestion. In the process of people taking urban rail transit, the convenience and efficiency of passengers' ticket purchase and entering and leaving the station is also a necessary link to ensure the overall efficiency. Therefore, the normal operation, low failure rate, and high maintenance efficiency of urban rail transit station equipment are necessary guarantees to improve the operation efficiency of urban rail transit. Using PHM technology to solve the problem of incomplete operation monitoring and low maintenance efficiency of urban rail transit station equipment is one of the methods worth further research and exploration. By analyzing the application status of PHM technology, combined with the concept of PHM technology, this paper puts forward the equipment fault prediction model and gives the overall scheme design according to the model framework.

Keywords Urban rail transit · PHM technology · Station equipment · Scheme design

---

H. Deng (✉) · H. Li · Y. Meng
School of Transportation, Nanning University, Nanning, China
e-mail: 549465788@qq.com

X. Cao
Guangxi Key Laboratory of International Join for China-ASEAN Comprehensive Transportation, Nanning University, Nanning, Guangxi, China

K. Subramaniyam and M. Paprzycki (eds.), *Smart Cities and the Internet of Things*, Internet of Things, https://doi.org/10.1007/978-3-032-08290-9_17

1 Introduction

As of now, the total number of urban rail transit operating stations in mainland China has reached 4,416. Based on the common single-channel gate equipment, the stations have a total of 96,086 gate channels (automatic ticket gate related channels), a total of 44,142 ticket machines (including cloud ticket machines), and a total of 15,672 semiautomatic ticket machines. As an important part of modern urban public transportation, the efficient and safe operation of urban rail transit is inseparable from the stable support of various equipment [1]. The automatic ticket vending machines (TVMs) and automatic ticket checking machines (AGMs) in station equipment directly serve passengers, and their operating conditions directly affect passengers' travel experience and the operating efficiency of the station [2]. At present, the means of operation, maintenance, and management of ticketing equipment in urban rail transit stations are relatively traditional and backward. They are all based on three dimensions: patrol inspection, regular maintenance, and fault repair to ensure the normal operation of the equipment. Daily patrol inspections are carried out by staff before daily operations. They conduct appearance inspections of automatic ticket vending machines and automatic ticket checking machines, including whether the outer shell is damaged, whether the display screen is functioning normally, and whether the buttons on the operation panel are sensitive. Regular maintenance is to carry out repair and maintenance operations according to different time periods [3]. During weekly inspections, maintenance personnel clean and lubricate the internal mechanical parts of the automatic ticket vending machines and automatic ticket checking machines to prevent dust accumulation from affecting the transmission effect and to extend the service life of the motor. During monthly inspections, the electrical systems of the equipment are inspected, including checking whether the line connections are loose, whether the output voltage of the power module is stable, etc. Quarterly and annual inspections are more comprehensive. In addition to in-depth inspections of the mechanical and electrical systems, the software system of the equipment is also be upgraded and optimized to improve the performance and function of the equipment. Fault repair means that when a malfunction occurs in the automatic ticket vending machine or automatic ticket checking machine, the station staff immediately notify the maintenance personnel, who use their experience and preestablished fault-handling procedures to troubleshoot and repair the problem in the shortest possible time. The current operation and maintenance management methods for ticket vending and checking equipment in urban rail transit stations are relatively backward and not intelligent enough. An intelligent operation and maintenance system based on dynamic monitoring and component life and health cycle prediction should be established to reduce equipment failure rates and operation and maintenance costs and to improve equipment operation efficiency [4, 5].

2 Design Requirements for Intelligent Operation and Maintenance System of Ticket Vending and Checking Equipment in Urban Rail Transit Stations

In order to ensure the normal operation of ticket vending and checking equipment in urban rail transit stations during line operation, avoid loopholes in traditional operation and maintenance methods, reduce the manpower and material costs of each link, improve equipment operation efficiency and reduce equipment failure rates, and promote the informatization and intelligent management of equipment [6], the design of the equipment intelligent operation and maintenance system should meet the following requirements:

- It can realize the real-time status monitoring function of the equipment. Using the Internet of Things technology, sensors are installed on key components of the automatic ticket vending machine (TVM) and the automatic ticket checking machine (AGM), such as the banknote recognition module, the gate drive motor, the ticket transmission device, etc., to collect equipment operation data in real time, including temperature, current, voltage, operating frequency, wear degree of mechanical parts, etc. Through the real-time data feedback, the operation status of the equipment is intuitively presented on the monitoring interface of the operation and maintenance system.
- It can realize the fault prediction and analysis function of the equipment. It uses backpropagation (BP) neural network algorithm and big data analysis to deeply mine the historical operation data and real-time collection data of the equipment. By building a fault prediction model, it can predict the possible fault type and time node of the equipment in advance.
- It can realize remote diagnosis and control functions. The system can provide remote operation functions for maintenance personnel. When the equipment fails, maintenance personnel can remotely connect to the faulty equipment through the intelligent operation and maintenance system without going to the site to perform diagnostic tests on the equipment. This includes remotely starting the equipment self-test program, adjusting equipment parameters, and updating software versions. For some simple faults, they can be repaired directly and remotely, reducing the time and cost of on-site maintenance.

3 Application of PHM Technology in Intelligent Operation and Maintenance System

PHM technology uses advanced sensor technology to obtain real-time status information and fault information of system operation [7]. Through equipment status monitoring and diagnosis, the equipment status and fault information can be obtained, and fault judgment can be made based on this, allowing its development trend to be

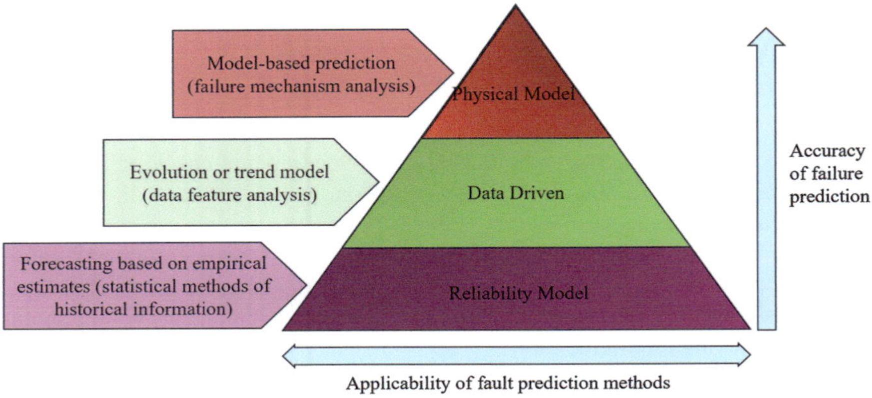

Fig. 1 Types of PHM fault prediction methods

predicted. Before the fault occurs, it can be determined whether to perform maintenance based on the equipment status [8]. Through large amounts of data analysis, the real-time status of the equipment can be obtained to provide early warning for the equipment and perform maintenance in advance. The status monitoring and diagnostic technology can also be used to obtain equipment status and fault information to determine equipment abnormalities, predict fault development trends, and decide whether to repair the equipment based on the equipment status before the fault occurs. Through a variety of network intelligent algorithms such as neural network algorithms, fuzzy logic algorithms, data fusion algorithms, and physical models, real-time monitoring of the entire system can be achieved, and system failures can be predicted before they occur [9]. From the perspective of modeling methods, the main fault prediction methods of PHM are mainly divided into: fault prediction methods based on reliability models, fault prediction methods based on data driven, and fault prediction methods based on physical models [10]. The types of fault prediction methods of PHM are shown in Fig. 1.

Taking PHM technology as the construction goal and combining the performance characteristics of automatic ticket vending machines and automatic ticket checking machines, it is more appropriate to choose a data-driven fault prediction method. The data-driven fault prediction process is shown in Fig. 2.

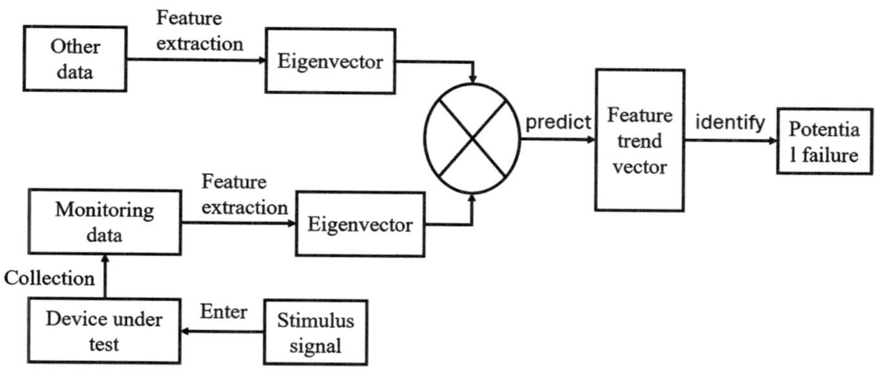

Fig. 2 Data-driven fault prediction process

4 System Scheme Design

4.1 System Architecture Design

In order to conduct a comprehensive analysis of the collected data, ensure the reasonable scheduling of each prediction model, and maintain the reasonable and standardized operation of the system, a hierarchical model is used for design. The system can be divided into six levels: data acquisition layer, data processing layer, state monitoring layer, fault diagnosis layer, fault prediction layer, and health decision layer. The PHM hierarchical structure is shown in Fig. 3.

- The function of the data acquisition layer is to collect data from various sensors on equipment such as ticket vending machines and ticket checking machines.
- The data processing layer is responsible for the realization of various system functions and is the core of the entire system. It uses certain methods to process the collected information, classify the data, and issue fault alarms. The system sets a normal threshold to compare the data feature value with the danger value to determine the degree of equipment failure.
- The function of the status monitoring layer is to monitor the collected data and conduct evaluation according to the preset thresholds, including the equipment's network interface, power module status, display interface, printing interface, ticket reader, cash handling device, ticket handling device, operation status display, stored-value ticket reader, system temperature, etc. The focus of monitoring is to compare the characteristic value of the data with the threshold to determine whether the equipment has failed and to assess the severity of the failure. If the real-time data is less than the warning value, it means that the system has no abnormal situation; if the real-time data is greater than the warning value and less than the danger value, it means that the system has failed and the automatic ticket vending machine needs to be inspected and repaired; if the real-time data

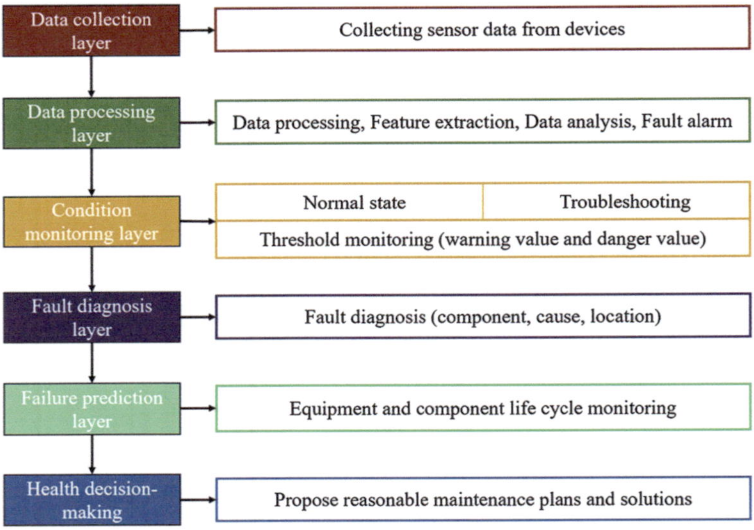

Fig. 3 PHM hierarchy diagram

is greater than the danger value, the automatic ticket vending machine needs to be repaired to avoid passenger flow backlog.

- The role of the fault diagnosis and fault prediction layer is to conduct big data analysis and comparison on the analysis results obtained by the status monitoring layer and confirm the fault information. When it is clear that the equipment is about to fail or has failed, the fault is diagnosed and the fault components, causes, locations, and severity are diagnosed.
- The role of the health decision-making layer is to conduct a reasonable and effective assessment of the comprehensive status of the equipment, analyze and judge the operating status and life of the equipment-related components, and provide reasonable warranty and maintenance opinions and solutions.

4.2 Data Acquisition Module Design

Relevant parameter information is collected through the sensor components of the automatic ticket vending machine and the automatic ticket checking machine, and the relevant data is converted into an effective form or format to classify, process, and pass the data to the status monitoring fault prediction system. The data collection workflow is shown in Fig. 4.

By selecting the core components of the equipment, the data dynamic monitoring and statistics are carried out, and the parameters of the monitoring components are set to the corresponding threshold. The system evaluates the operating status and life of the corresponding components according to the comparison between the dynamic

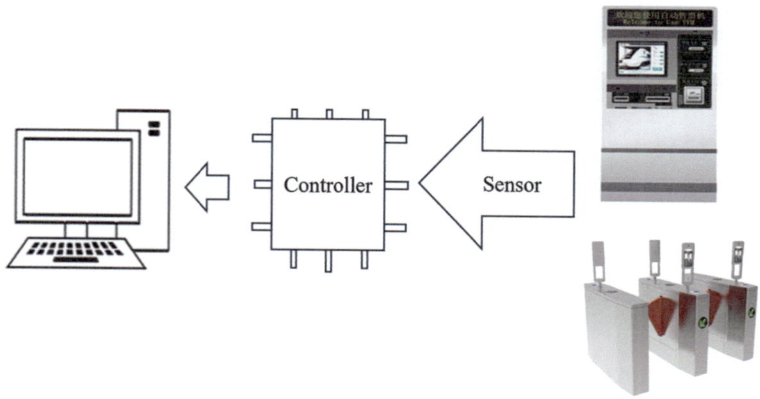

Fig. 4 Data collection workflow diagram

data and the corresponding threshold. The selection and parameters of components are shown in Table 1.

4.3 Data Fusion Module Design

In order to avoid a large number of errors in the data during the data collection process, data fusion is used to classify the collected data and filter out useless information from the data. If a new fault is found during monitoring, the new error must be combined with the previous old error, and the new fault category must be entered into the system. After processing, the results become more accurate and effective. The data fusion model is shown in Fig. 5.

4.4 Management System Platform Construction

The final presentation of the management platform is the display of the system's judgment results. For the fault information obtained after analysis, the diagnosis results must first be exported and displayed. If the judgment result exceeds the danger value, the system will start the alarm monitoring for fault prediction and provide reasonable operation and maintenance solutions. The health management system platform is shown in Fig. 6.

Table 1 Device component running parameters and threshold settings

Equipment part name	Normal operating parameters and ranges of each module					Monitoring threshold setting		
	Operating temperature range (°C)	Operating humidity range	Voltage value (V)	Current value (mA)	Power consumption (W)	Voltage threshold (V)	Current threshold (mA)	Power consumption threshold (W)
Reader	0–60	0–95%	DC12 ± 0.5	700	10	DC11.5≤, ≥DC12.5	≥700	≥10
Distribution module	0–60	0–95%	DC24 ± 0.5	2000	50	DC23.5≤, ≥DC24.5	≥2000	≥50
Passenger display touch screen	0–60	0–95%	DC12 ± 10%	85	3	DC1O.8≤, ≥DC13.2	≥85	≥3
Coin module	0–60	0–95%	DC24 ± 0.5	1000	25	DC23.5≤, ≥DC24.5	≥1000	≥25
Banknote module	0–60	0–95%	DC24 ± 10%	1000	25	DC21.6≤, ≥DC26.4	≥1000	≥25
Integrated communication controller	0–60	0–95%	DC24 ± 10%	800	20	DC21.6≤, ≥DC26.4	≥800	≥20
Industrial personal computer	0–60	0–95%	DC12 ± 10%	1500	40	DC1O.8≤, ≥DC13.2	≥1500	≥40
PCM movement control board	0–60	0–95%	AC36 ± 5%	3000	100	AC34.2≤, ≥AC37.8	≥3000	≥100
Door motor	0–60	0–95%	DC48 ± 0.5	2000	96	DC≤47.7, ≥DC48.5	≥2000	≥96

(continued)

Table 1 (continued)

Equipment part name	Normal operating parameters and ranges of each module					Monitoring threshold setting		
	Operating temperature range (°C)	Operating humidity range	Voltage value (V)	Current value (mA)	Power consumption (W)	Voltage threshold (V)	Current threshold (mA)	Power consumption threshold (W)
Direction indicator	0–60	0–95%	DC24 ± 0.5	500	10	DC23.5≤, ≥DC24.5	≥500	≥10
Passenger display screen	0–60	0–95%	DC12 ± 0.5	450	8	DC11.5≤, ≥DC12.5	≥450	≥8

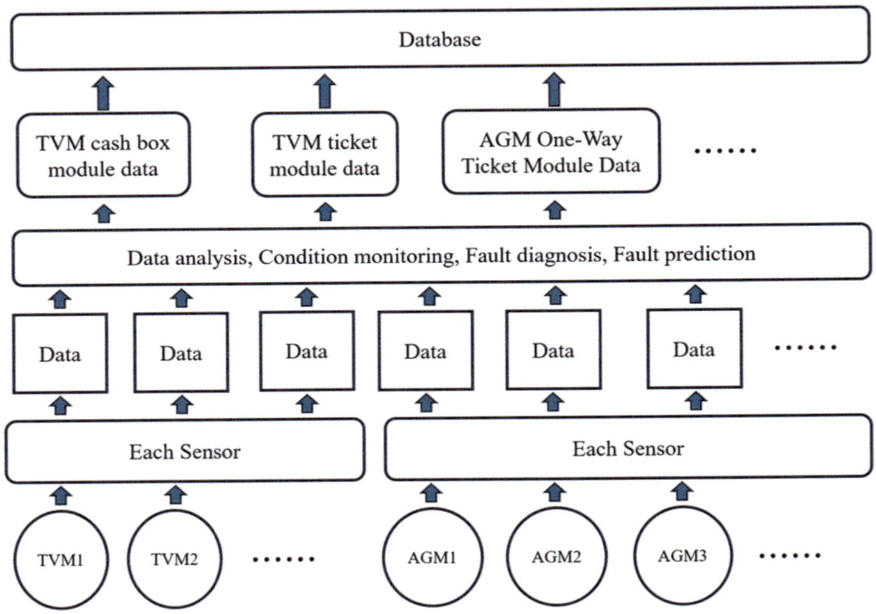

Fig. 5 Data fusion model diagram

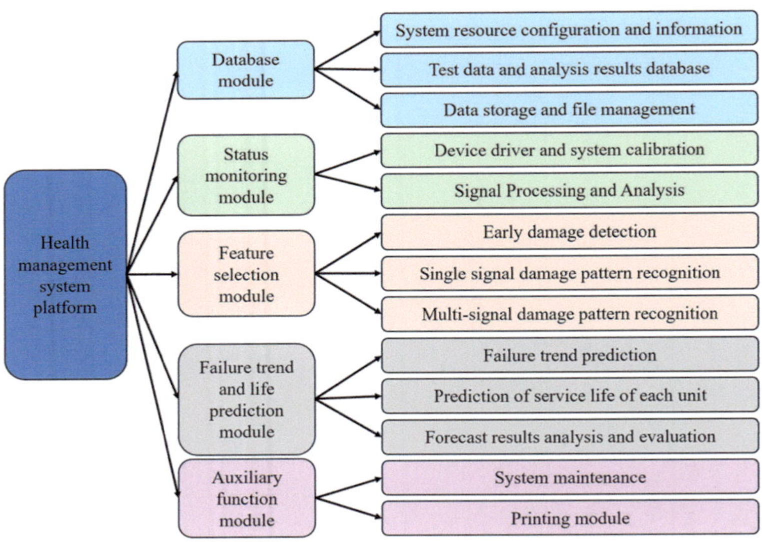

Fig. 6 Health management system platform

5 Conclusions

This design focuses on the operation and maintenance problems of ticket vending and checking equipment in urban rail transit stations and builds an intelligent operation and maintenance system based on PHM technology. The system integrates functional modules such as multisource data collection, data analysis and processing, fault prediction, and intelligent decision-making, aiming to improve the efficiency and reliability of equipment operation and maintenance. Through real-time monitoring and precise analysis of the equipment's operating status, potential fault hazards can be detected in advance, the traditional post-maintenance mode can be changed, preventive maintenance can be achieved, the equipment failure rate can be greatly reduced, and the efficiency and stability of rail transit operations can be guaranteed. Intelligent operation and maintenance system can significantly reduce operation and maintenance costs by improving efficiency, optimizing resource allocation, and predicting faults in advance. In terms of labor costs, material costs, time costs, etc., it can generally save about 8% of the comprehensive cost. With the continuous advancement of technology and the expansion of application scenarios, the intelligent operation and maintenance system can also strengthen the seamless connection with the rail transit operation and management system, closely linking equipment operation and maintenance information with train dispatching, passenger flow management, and other links. This provides more powerful support for the overall efficient operation of urban rail transit and helps the urban rail transit industry move toward a smart and sustainable direction.

Acknowledgements This project is supported by the Nanning Scientific Research and Technology Development Plan Project Artificial Intelligence for IT Operations and Health Monitoring Platform for Transport Infrastructure, No. 20223230 and the Guangxi Science and Technology Base and Talent Project: Construction of Guangxi Transportation New Technology Transfer Center Platform, Guike AD23026029.

References

1. Qi Xing. (2022). On the basic principles and safeguard measures of urban public transportation management. China Storage and Transportation (10), 81-82. https://doi.org/10.16301/j.cnki.cn12-1204/f.2022.10.050.
2. Jian, Z.: Composition and implementation effect evaluation of the information management system for outsourced maintenance of electromechanical equipment in urban rail transit stations. Urban Rail Transit Research **06**, 152–156 (2017). https://doi.org/10.16037/j.1007-869x.2017.06.034
3. Cen Yixing. (2021). A brief analysis of the new concept of urban rail transit equipment maintenance management based on big data. China Equipment Engineering (21), 43–45. CNKI:SUN:SBGL.0.2021-21-024.
4. Zheng, X.D. (2021). Research on equipment life cycle management and fault prediction (Master's degree thesis, Southwest University of Science and Technology). https://kns.cnki.net/KCMS/detail/detail.aspx?dbname=CMFD202102&filename=1021638391.nh

5. Zhao Shuai. (2018). Research on key technologies for equipment remaining life prediction based on data drive (PhD dissertation, Northwestern Polytechnical University). https://kns.cnki.net/KCMS/detail/detail.aspx?dbname=CDFDLAST2020&filename=1020706635.nh
6. Zeng Jinglin. (2025). New concept of urban rail transit equipment maintenance management based on big data. Transport Manager World (03), 1–3. CNKI:SUN:YSGL.0.2025-03-001.
7. Dai Rui. (2024). Research and application of PHM technology for metro electromechanical equipment. China Machinery (18), 67–70. CNKI: SUN:HUJX.0.2024-18-016.
8. Zhiyuan, L., Jiangming, D., Yao, S., Yingmou, Z., Zhong, H.: Current status and development trend of PHM technology for rail transit equipment. Electric Locomotives and Urban Rail Vehicles **03**, 8–16 (2024). https://doi.org/10.16212/j.cnki.1672-1187.2024.03.002
9. Jiang Feng. (2024). Research on big data analysis and decision support technology in power system automation. Digital Communication World (11), 107–109. CNKI:SUN:SZTJ.0.2024-11-032.
10. Wu Huali, Jiang Yanan & Chen Jundong. (2021). Framework design of intelligent operation and maintenance system for urban rail transit based on PHM technology. Intelligent Building and Smart City (12), 152-154. https://doi.org/10.13655/j.cnki.ibci.2021.12.062.

The Design of a Virtual–Real Hybrid Simulation System for Airborne Weather Radar

Shiwei Zhao, Menglong Shao, Feilong Huang, and Xibo Cao

Abstract To address equipment limitations, environmental constraints, and potential safety risks in the teaching and training of airborne weather radar systems, a virtual simulation technology combined with physical components has been designed and implemented. This integrated system replicates both the structure and functions of actual airborne weather radar systems. Adhering to the ARINC 429 and ARINC 708 technical specifications, software programs for human–machine interaction were developed on the Visual Studio platform. This system can replicate the operational testing and system testing of real airborne weather radars, and through component testing interfaces, it enables in-depth maintenance testing of radar system accessories. Practice has proven that the system is safe, reliable, and cost-effective. The provided radar system maintenance engineering practice scenarios effectively enhance students' competencies for relevant positions, making them significantly valuable in engineering practice teaching related to aircraft maintenance.

Keywords Airborne weather radar · Virtual–real integration · Simulation validation · In-depth maintenance testing

1 Introduction

The airborne weather radar system is a crucial component of civil aviation aircraft, enabling continuous monitoring of hazardous weather conditions and providing real-time meteorological information during flight to ensure flight safety. According to the latest provisions of Annex 6 to the Convention on International Civil Aviation—Operation of Aircraft, Part I: International Commercial Air Transport—airplanes and fixed-wing aircraft operating at night and/or under instrument meteorological

S. Zhao
Engineering Training Center, Civil Aviation University of China, Dongli, Tianjin, China

M. Shao (✉) · F. Huang · X. Cao
School of Electronic Information and Automation, Civil Aviation University of China, Dongli, Tianjin, China
e-mail: 2522644040@qq.com

K. Subramaniyam and M. Paprzycki (eds.), *Smart Cities and the Internet of Things*, Internet of Things, https://doi.org/10.1007/978-3-032-08290-9_18

conditions (IMC) must be equipped with an airborne weather radar system. National airworthiness certification regulations further specify detailed requirements for such systems [1].

Literature research indicates that although numerous institutions offer aviation engineering programs, the teaching of airborne weather radar systems remains predominantly theoretical. As a result, there is insufficient practical training. Practical instruction is constrained by equipment availability and environmental limitations, resulting in a disconnect between the training content and hands-on experience. Consequently, students struggle to develop a comprehensive understanding of radar system operation and to enhance their maintenance competencies. Moreover, the high acquisition and operational costs of actual airborne weather radar systems, along with potential safety hazards associated with microwave radiation, make them unsuitable for large-scale training without modifications.

This study integrates virtual simulation and physical components to develop an airborne weather radar simulation system; it encompasses both hardware circuit design and software implementation. The system features an intuitive human–machine interaction interface and facilitates an integrated "observe-check-listen-operate-analyze" teaching approach. By optimizing the avionics maintenance training model, it enhances students' immersive learning experience and improves overall teaching effectiveness.

2 Requirements Analysis of the Airborne Weather Radar Simulation System

2.1 *Analysis of Teaching Function Requirements*

The functional requirements of the airborne weather radar simulation system originate from its maintenance tasks in various scenarios, including real-world line maintenance, hangar inspections, and component repairs. To address the equipment limitations, environmental constraints, and potential safety risks in the teaching and training of airborne weather radar systems and to enhance students' ability to solve complex civil aviation maintenance engineering problems, the simulation system should meet the following requirements:

System completeness: The system should ensure both structural and functional completeness of the airborne weather radar system, providing comprehensive training capabilities.

Engineering practicality: implement radar echo images and system mode indications under different operating modes based on ARINC 708 data.

Human–machine interaction: provide an intuitive interactive interface that supports user operation of the system, enabling functionalities such as power control, radar control panel operation, electronic flight instrument system (EFIS) control panel operation, and testing.
Operational safety: mitigate the risks of microwave radiation associated with real equipment operation through virtual simulation technology. The system should reference aircraft maintenance manuals to ensure safety in engineering practice.

2.2 Overall Architecture Design of the Simulation System

To meet the maintenance requirements and educational objectives of the airborne weather radar system, the simulation system should assist maintenance engineers in mastering the functionality and interrelationships of system components, interpreting system display information, and developing the ability to identify risk sources, locate faulty components, and formulate troubleshooting plans, thereby enhancing job competency.

Based on the analysis of system functional requirements, this study designs the overall architecture of the simulation system [2], which is structured into three layers: user layer, data layer, and hardware layer, as illustrated in Fig. 1.

The user layer is responsible for direct interaction with the user, providing an interface for display and operation. It enables users to control the system and receive

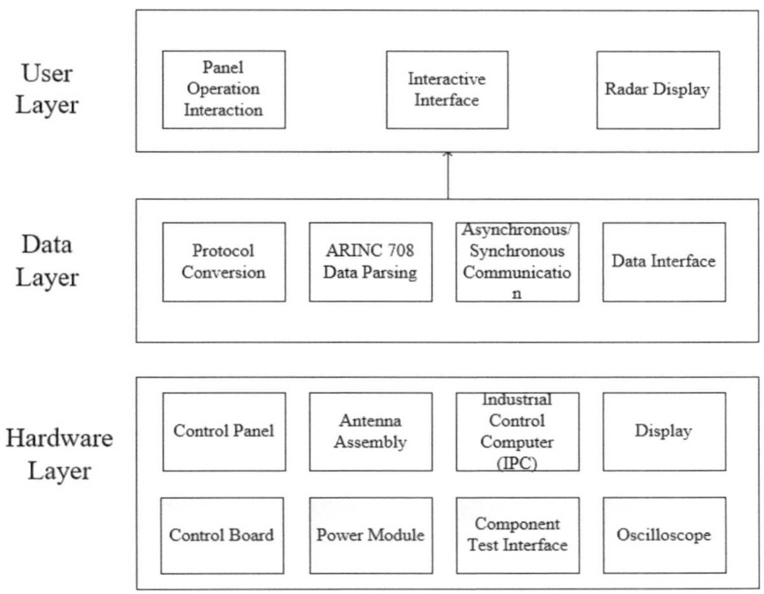

Fig. 1 Overall architecture design of the simulation system

feedback through radar data visualization, human–machine interaction interface, and control panel operations.

The data layer handles data transmission, parsing, and protocol conversion between different components, enabling their interconnection. It facilitates the generation of radar system data displays and operational status visualization on the monitor.

The hardware layer consists of physical devices and underlying hardware resources, providing the necessary support for system operation while ensuring efficient resource allocation. It primarily includes control box, antenna components, industrial control computer (IPC), display, control board, power module, oscilloscope, and component testing interfaces.

To enhance the realism of the simulation system, both the control box and the antenna components are implemented as actual hardware. However, due to the high cost of transceivers and their limited interaction with other system components during training, they fail to effectively simulate the integrated operation of the system in a real-world environment. Additionally, the potential safety risk of the radar system arises from the microwave radiation emitted by the transceiver. Considering both practicality and safety, the study uses an industrial control computer (IPC) as a substitute.

3 Module Design of the Airborne Weather Radar Simulation System

Based on the teaching requirements analysis and overall architecture design, the simulation system is structured around three core modules: radar display module, antenna servo control module, and control panel function module. The radar display module provides radar echo images and serves as a human–machine interaction interface for users. The antenna servo control module simulates the scanning motion of a real radar antenna, implementing the actual radar control logic and operational modes. The control panel function module replicates the control logic of a real radar system, ensuring that the simulation system's operational experience aligns with actual equipment.

Additionally, each functional module is equipped with test interfaces for data acquisition and in-depth component maintenance testing, facilitating system analysis and fault diagnosis.

3.1 The Radar Display Module

This paper generates radar images in different modes based on ARINC 708 data. The data is produced by the antenna emitting microwave signals, which are reflected by raindrops or cloud layers, then received by the transceiver, and processed to form the radar image. The ARINC 708 meteorological data is continuously output in data words, each composed of four main parts:

1. Control and mode settings: used to express the radar's operating mode.
2. Operating status: used to display the radar's current working status, such as ANT, TILT, etc.
3. Scan angle: used to control and display the radar's scan angle.
4. Radar data: used for image display, adopting the RGB color model. Starting from bit 65, every 3 bits represent one pixel color. Each data word contains 512 pixels of data, which is output in serial format. The data output sequence corresponds to the pixel positions along the scan line, with different colors reflecting varying radar echo intensities.

The radar antenna's scanning angle range is 180° horizontally, resulting in a semicircular sector as the actual radar image display area (as shown in Fig. 2). In ARINC 708 data, bit 52 is the least significant bit (LSB), and bit 63 is the most significant bit (MSB), allowing for a maximum representation of angles ranging from 0° to 360°. The scanning angle in the data word and the actual radius used for rendering on the screen are determined by Eqs. (1) and (2).

$$\theta = \sum_{i=52}^{63} b_i \times \left(180 \times 2^{-(63-i)}\right) \tag{1}$$

$$\rho = \frac{Dx}{512} \times R \tag{2}$$

In the equation:

- b_i represents the binary value from bit 52 to bit 63.
- θ denotes the scanning angle.
- D represents the actual distance.
- D_x is the distance of a specific pixel from the image center during rendering, with a range of 0–511.

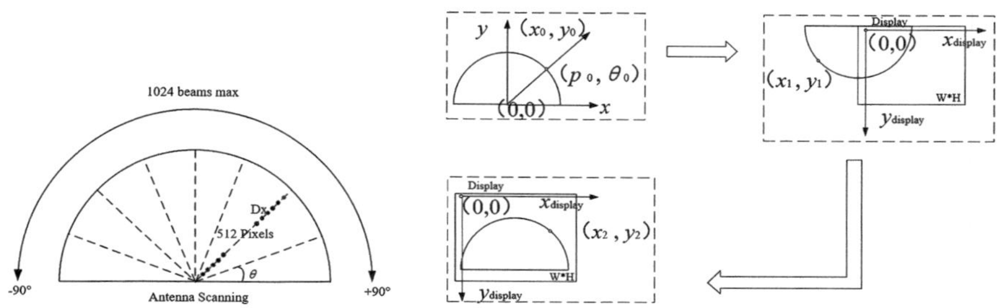

Fig. 2 Radar image display and coordinate transformation

- R is the radius of the image display area.

ARINC 708 meteorological radar data is defined in a polar coordinate system, whereas standard displays use a Cartesian coordinate system, with the top-left corner as the default origin (0, 0). To correctly display the radar image, a coordinate and position transformation is required, aligning the image origin to the center-bottom of the display window. For a display resolution of W × H, the center position is (W/2, H). Figure 2 illustrates the coordinate transformation process, which can be computed using the matrix defined in Eq. (3).

$$\begin{pmatrix} X_2 \\ Y_2 \end{pmatrix} = \begin{pmatrix} W/2 + \rho_0 \cdot \cos\theta_0 \\ H - \rho_0 \cdot \sin\theta_0 \end{pmatrix} \tag{3}$$

3.2 The Antenna Servo Control Module

To enhance the realism of the simulation system and replicate actual flight conditions, the system incorporates the WMA-701 radar antenna assembly. Since the industrial control computer (IPC) cannot directly drive the antenna, a PCI-8164 motion control card is used in conjunction with the IPC to construct the antenna motion control system [3, 4], as shown in Fig. 3.

The stepper motor within the antenna mount requires differential square-wave pulse signals for operation. Therefore, in this study, the motion control card is configured to A/B phase pulse output mode to generate differential square-wave pulse signals consistent with the transceiver unit. By controlling the pulse frequency and count, the antenna's movement angle and speed are precisely regulated.

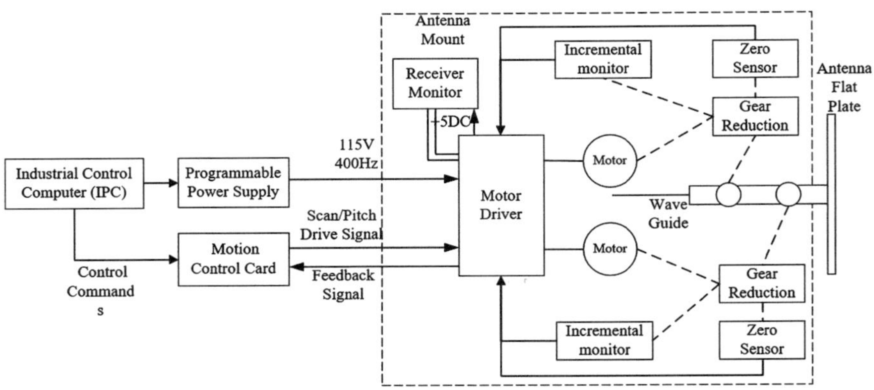

Fig. 3 Antenna motion control system block diagram

Based on the motion characteristics of the real system, this study designs a control software program using the C# language. The program utilizes the DllImport attribute to declare external functions, enabling the invocation of dynamic link libraries (DLLs) for controlling the motion control card.

The workflow of the antenna motion control software is as follows:

1. Initialize the motion control card, configuring the pulse output mode, feedback source, and other parameters.
2. Reset commands, position, and error counters to ensure the system starts in an initial state.
3. Implement multithreaded motion control, enabling horizontal scanning and pitch adjustments, ensuring consistency with the real system's motion characteristics.
4. The reduction ratio between the flat-panel antenna and the stepper motor is 180:1. To meet the requirement of 45° per second for the antenna's operating speed, the maximum speed (MaxVel) is set to 180 pulses per second (PPS) in the software.

3.3 The Control Panel Function Module

In the simulation system, the radar control panel serves not only as an interactive control device but also as a data communication interface. To ensure that the system's control operations align with those of a real radar system, this study incorporates the WCP-701 radar control panel, which communicates with the transceiver unit via the ARINC 429 bus. However, since the industrial control computer (IPC) cannot directly recognize the differential signals of the ARINC 429 bus, a protocol conversion is required to translate ARINC 429 signals into a general bus format recognizable by the IPC.

To meet the ARINC 429 protocol conversion requirements and ensure stable system communication, this study employs the PCI-7102 board for protocol conversion, as illustrated in Fig. 4.

The control panel commands are transmitted via the ARINC 429 bus to the interface circuit of the board, where signal level conversion is performed. The converted data is then processed by the field-programmable gate array (FPGA), which translates it into PCI bus data and forwards it to the industrial control computer (IPC). This enables the IPC to recognize the radar control panel commands, establishing the foundation for interconnection and control among components within the radar simulation system.

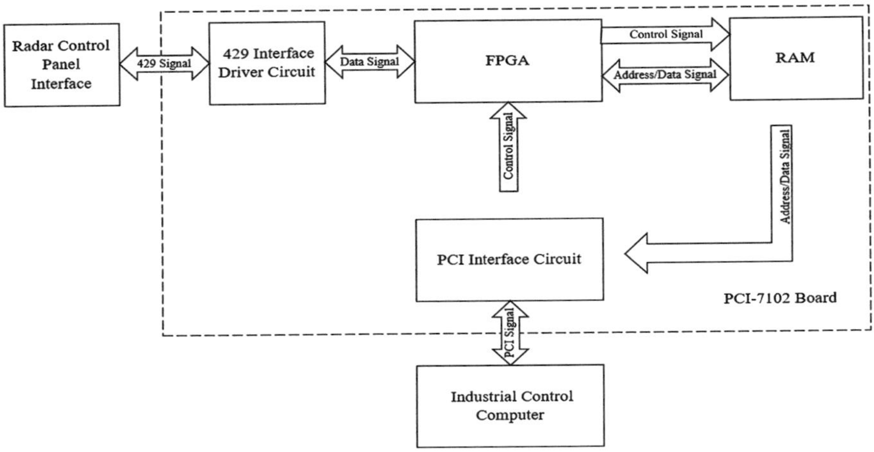

Fig. 4 A 429 board protocol conversion principle block diagram

4 Implementation of the Simulation System

4.1 Structure and Functional Completeness of the Simulation System

According to the teaching requirements of the simulation system and following the overall design architecture, the physical implementation of the airborne weather radar simulation system has been completed. The system primarily consists of a radar antenna assembly, control panel, industrial control computer (IPC), and industrial display, as shown in Fig. 5.

The simulation system replicates the hardware structure of the real system by incorporating identical components. Additionally, it implements key functions, including self-test, weather detection, turbulence monitoring, and mapping. This corresponding design enables the system to reproduce the control logic and functionalities of the real system under various scenarios.

To ensure the prevention of potential harm to personnel or equipment during system operation, an emergency stop switch is designed to promptly halt the system's operational state in case of emergencies.

By analyzing the weather radar data from the ARINC 708 standard, the simulation system's software design adopts multithreading technology to ensure low-latency radar image generation. Additionally, the system allows users to switch operational modes, adjust gain settings, and perform cyclic antenna scanning through the radar control panel. A data binding mechanism is implemented to ensure that user inputs are immediately reflected in the radar images, providing an interactive experience consistent with that of an actual radar system. Furthermore, the system incorporates a continuous update mechanism for antenna position and scanning angle, ensuring

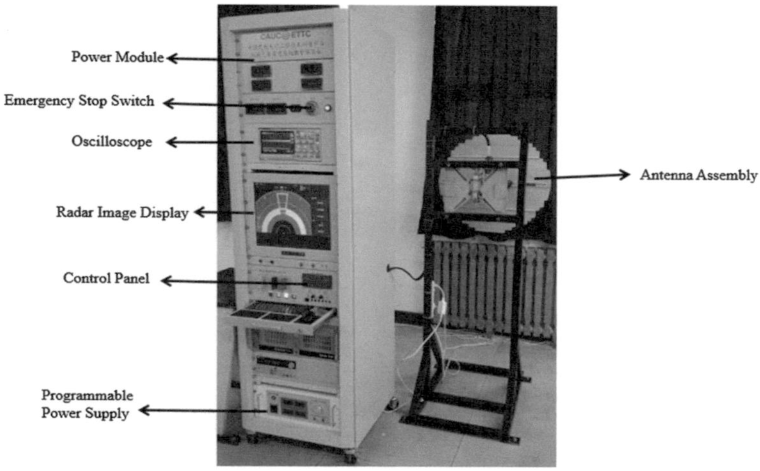

Fig. 5 Physical diagram of the simulation system

that the generated radar images remain highly synchronized with the actual radar operation, as shown in Fig. 6.

In the radar images shown in Fig. 6, Fig. 6a indicates that the radar operates in "WX" mode, with an antenna tilt angle of 6°, a gain mode of VAR, and a detection range of 80 nautical miles (with the displayed value being half of the actual value). Figure 6b shows that the radar operates in "WX + T" mode, with an antenna tilt angle of 4°, a gain mode of VAR, and a detection range of 40 nautical miles. Through comparative analysis with actual radar images, the simulation system effectively replicates the functionality of airborne weather radar. This replication meets the requirements for real-time data processing and interaction, while also providing reliable support for subsequent system testing and educational applications.

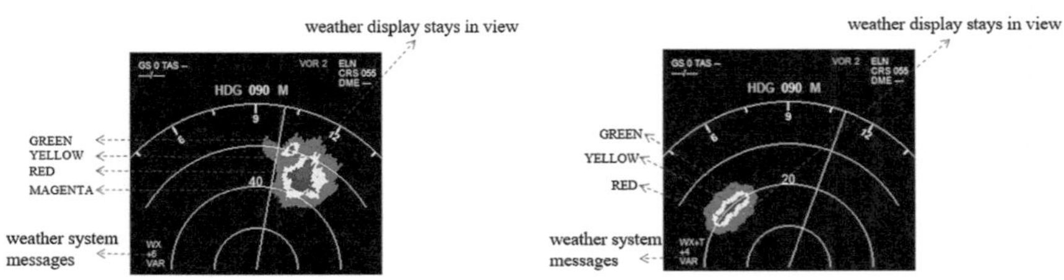

a.WX Mode Radar Display b.WX+T Mode Radar Display

Fig. 6 Typical radar images

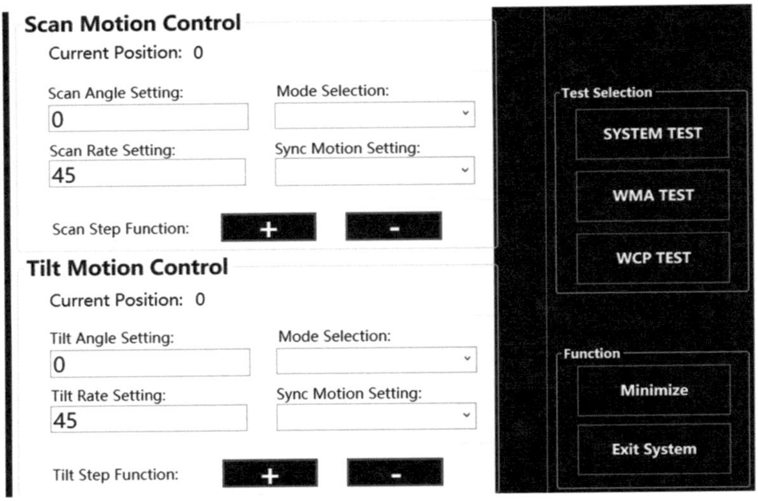

Fig. 7 WMA TEST human–machine interface (HMI)

4.2 High-Fidelity Student–System Interaction

To meet further teaching requirements, a motion control test program for the antenna assembly was developed based on the Component Maintenance Manual (CMM). This program includes three functional modules: antenna scan control, antenna pitch control, and system master control, and is equipped with a human–machine interface (HMI) [5].

Each module provides intuitive parameter input fields and mode selection functions, allowing users to adjust motion parameters such as scan range and scan rate. Additionally, the system supports data binding functionality, ensuring real-time feedback on parameter changes and dynamically updating antenna position information on the interface, as shown in Fig. 7.

5 Conclusion

This study focuses on the training functional requirements of the airborne weather radar system. It details the design and implementation of each functional module within the simulation system and successfully completes the development of the airborne weather radar simulation system. Through multiple tests and practical teaching applications, the system has demonstrated high stability and ease of use.

Student feedback indicates that the interactive design of the simulation system is well structured. This design facilitates a deeper understanding of the working principles of the airborne weather radar system. Additionally, by integrating continuous airworthiness maintenance technical documents and airborne equipment technical specifications, the system significantly enhances trainees' fault analysis and troubleshooting skills in line maintenance, periodic inspections, and component maintenance. The research results confirm that this simulation system can effectively improve students' job competency, providing a reliable practical platform for the teaching and training of airborne weather radar systems.

Acknowledgements This study is one of the phase achievements of the National-Level Second Batch New Engineering Research and Practice Project, titled "Exploration and Practice of Innovation and Entrepreneurship Capability Development for New Engineering Talents in the New Civil Aviation Industry" (Project No. E-CXCYYR20200910).

References

1. Li Hai, Zhou Anyu, Chen Nannan. "Simulation and Verification of Thunderstorm Echoes in Airborne Dual-Polarization Weather Radar" [J]. Signal Processing, 2024, 40(09): 1728–1737.
2. Cong, C., Yang, J., Xuan, W., et al.: Development of a Virtual Simulation Practice Teaching Platform for Aircraft Systems" [J]. Laboratory Research and Exploration **36**(04), 83–86 (2017)
3. Miki Shichijo, Ryuichi Oguro. "Model following control with a sliding mode adjustment for a positioning system" [C]. Proceedings of SICE-ICASE International Joint Conference. Busan: SICE, ICASE, 2006: 292–295.
4. Subudhi, B.: MORRIS A S "Soft computing methods applied to the control of a flexible robot manipulator" [J]. IEEE Trans. Neural Networks **12**(1), 54–67 (2001)
5. Chakraborty, S., Aithal, P.S.: MVVM Demonstration Using C# WPF[J]. International Journal of Applied Engineering and Management Letters (IJAEML) **7**(1), 1–14 (2023)

Research and Development of a BIM+IoT--Based Digital Command and Decision Platform for Construction Engineering

Junyi Li, Xiaodan Ma, Hang Yu, Yansong Wang, Xuejuan Dou, Pengzhou Wang, and Yang Wang

Abstract Traditional construction projects are characterized by large scale, long duration, and multiple stakeholders, leading to issues such as inefficient management and low operational efficiency. Under the national digital transformation strategy, the construction industry must embrace digital transformation. By integrating artificial intelligence (AI), Internet of Things (IoT), and geographic information system (GIS) technologies, this study develops a command and decision-making system tailored for construction project management. The platform employs multidimensional and multilevel analysis through intelligent algorithms to enable data-driven decision-making. It achieves integrated control over key construction elements such as personnel, machinery, materials, methods, environment, quality, safety, and progress, while facilitating precise supervision and instruction delivery to project teams. Pilot applications validate the platform's effectiveness, offering valuable insights for digital transformation in similar projects.

Keywords Command and decision platform · Staff management · Machinery management · Material management · Technical management · Environmental management · Quality management · Safety management · Progress management

J. Li · X. Ma · Y. Wang
Beijing Zhongjian Jingdong Real Estate Co., Ltd., Beijing, China

H. Yu · X. Dou
China Construction Equation Investment Development Group Co., Ltd., Beijing, China

P. Wang
Northwest Branch of China Construction Fourth Engineering Bureau Co., Ltd., Shaanxi, China

Y. Wang (✉)
China Construction Engineering Industry Technology Research Institute Co., Ltd., Beijing, China
e-mail: 2799939032@qq.com

China Construction Research Institute Intelligent Technology (Beijing) Co., Ltd., Beijing, China

K. Subramaniyam and M. Paprzycki (eds.), *Smart Cities and the Internet of Things*, Internet of Things, https://doi.org/10.1007/978-3-032-08290-9_19

1 Introduction

The construction industry, as a pillar of the national economy, plays a vital role in supporting China's economic development and employment stability. However, traditional construction practices have long been plagued by multiple challenges including labor-intensive operations, extensive management practices, overcapacity, low profitability, and operational inefficiency, all of which severely constrain high-quality development in the sector [1, 2]. Historically, the performance growth of construction enterprises primarily relied on macroeconomic policies and fixed-asset investments, manifesting as a singular development model dependent on continuous input of human and production resources for scale expansion. This has resulted in low gross profit margins and weak profitability, with industry profit rates persistently remaining at 2–4%—far below other sectors [3].

Moreover, the traditional construction industry remains high risk, characterized by complex safety challenges and frequent occurrence of severe accidents [4, 5]. Current economic deceleration has further impacted the sector, with annual growth of construction output value slowing from double digits to 6%, coupled with reduced growth in building construction areas. The era of widespread infrastructure boom has ended, and the industry's development logic has undergone fundamental changes. In this context, digital transformation for quality and efficiency enhancement has become the critical focus. The sector now faces urgent demands to transition from scale-driven expansion to innovation-driven, refined management models to achieve sustainable development.

Under these circumstances, traditional project management approaches struggle to maintain precise control over overall operations. While consuming enormous management costs, they often suffer from information gaps that lead to inefficiencies and even quality control failures [6, 7]. Project management entities must prioritize developing a BIM-GIS-IoT integrated digital command and decision-making platform—purpose-built for large-scale construction projects—to address these systemic challenges [8].

To address these needs, we have developed a BIM+IoT--based command and decision-making system featuring:

- A comprehensive data mapping framework that aggregates real-time data from smart site systems across project departments and extracts critical metrics.
- A multi-intelligence algorithm engine enabling multidimensional, multitiered data analysis.

Leveraging data-driven decision-making as the core methodology, the system achieves: integrated control over key construction elements (human resources, machinery, materials, methods, environment, quality, safety, and progress) at the command center level and directive communication and precise oversight of all project departments. This architecture ensures unified project governance while enhancing operational transparency and execution efficiency.

2 Platform Architecture

The platform adopts an integrated BIM + GIS + IoT technological framework and employs a multitiered microservices architecture with "cloud-edge-device" collaboration. The front-end utilizes a WebGL-based 3D engine to create lightweight web visualization interfaces, enabling online BIM model viewing and real-time IoT device status monitoring. The mobile terminal integrates augmented reality (AR) technology to achieve virtual–physical interaction between the digital twin and physical construction sites. The back-end leverages a hybrid cloud architecture with Kubernetes containerized deployment, utilizing elastic computing resource pools to enable high-concurrency data processing capabilities that support multiparty collaboration involving thousands of users.

The application layer establishes a comprehensive digital management system covering all construction elements through seven core functional modules:

Personnel Management: Implements full-staff digital authentication via facial recognition, ultra-wideband (UWB) positioning, and blockchain technology, generating real-time workforce heatmaps and dynamically optimizing crew allocation strategies. Integrated with virtual reality (VR) safety training systems, it elevates certification pass rates to 95%.

Machinery Management: Utilizes vibration sensors and digital twin technology to enable predictive maintenance for over 200 types of construction equipment, reducing idle rates through equipment utilization analysis models.

Technical Management: Develops a BIM-based standard process library incorporating multiple construction node specifications. An AR-assisted disclosure system achieves 3D visualization of construction methods, shortening design change response times to under 2 h.

Environmental Management: Deploys multisource sensor networks to monitor 12 environmental parameters (including $PM_{2.5}$ and noise levels). An intelligent spray control system activated by meteorological data reduces dust pollution incidents by 75%.

Schedule Management: Innovatively applies BIM+IoT- integration for 4D schedule simulation, automatically detecting workflow conflicts through construction sequence analysis.

Quality Management: Constructs an AI inspection system using computer vision (CV) to automatically identify multiple defect types, achieving 98.6% closed loop resolution rates for quality issues.

Safety Management: Develops dynamic risk assessment matrices that monitor numerous hazard sources in real time via UWB positioning and CV algorithms, enabling 30-min emergency response from detection to resolution.

As shown in Fig. 1, all modules achieve deep data interoperability through microservices architecture, supporting cross-organizational collaboration among engineering, procurement, and construction (EPC) contractors, subcontractors, and supervision teams. Compared with traditional management approaches, the platform

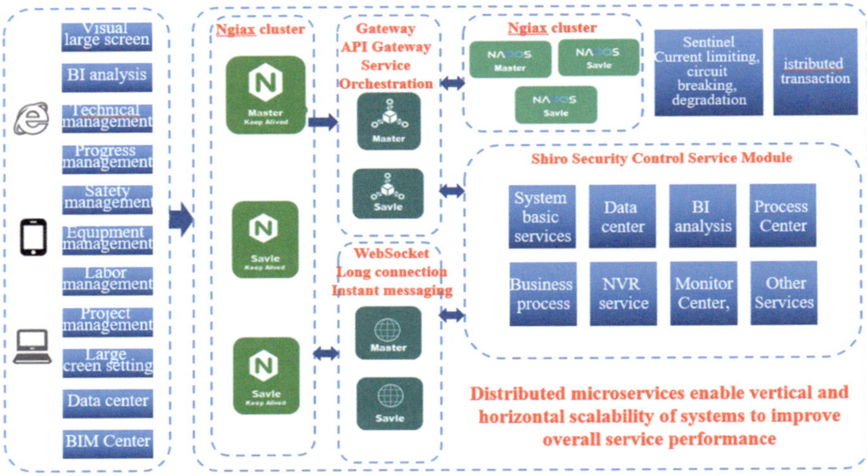

Fig. 1 Platform architecture design

demonstrates 3x higher decision-making efficiency. In a metro hub project, its multi-dimensional data analysis capabilities provided 23 early warnings for critical risks, contributing to the project's national engineering excellence award, marking a new era of "digital twin + intelligent decision-making" in construction management [9].

3 Platform Functions

Develop a command and decision-making system based on BIM+GIS. A comprehensive data mapping framework is developed to achieve real-time aggregation of smart construction site system data from various project departments and extract critical metrics. A multi-intelligent algorithm engine is created to enable multidimensional, multitiered data analysis, with data-driven decision-making as the core approach. This system empowers the command center to implement holistic control over all construction elements (human, machinery, materials, methods, environment, quality, safety, and schedule) while enabling effective directive communication and precise supervision of project departments.

The functional modules of this platform primarily include BIM model management, IoT device management, and digital command and decision-making. The BIM model management module enables 3D visual representation of building information. The IoT device management module facilitates real-time monitoring and control of connected devices. The digital command and decision-making module leverages data analysis and decision support to provide a scientific foundation for project management. These modules collaborate synergistically, forming a comprehensive digital command and decision-making platform tailored for construction engineering.

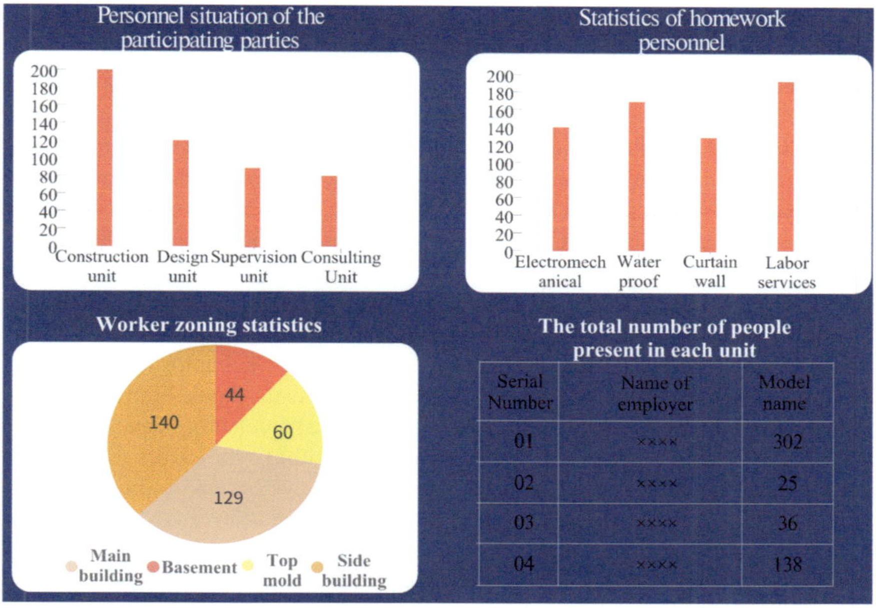

Fig. 2 Personnel management function diagram

3.1 Personnel Management

As illustrated in Fig. 2, in compliance with the labor real-name system management requirements of the construction industry, the system aggregates and analyzes data on organizational structures, staffing allocations, and attendance rates of project management and supervisory personnel. It also visualizes, statistically analyzes, and evaluates records and performance metrics of subcontractors, work teams, and laborers, while issuing real-time alerts for noncompliant activities. The personnel management module encompasses seven core functionalities: Personnel Profiles, Subcontractor Management, Work Team Management, Attendance Management, Blacklist Management, Safety Behavior Star Program (recognizing safe practices), and Credit Evaluation. This integrated framework ensures systematic oversight, enhances compliance, and strengthens risk control across all labor management processes.

3.2 Machinery Management

In compliance with national regulatory requirements, the project equipment management module deeply integrates IoT sensing technologies and intelligent algorithms to establish a full-chain digital management system. As shown in Fig. 3, the system interfaces with on-site smart access control systems, Global Positioning System (GPS) devices, and equipment radio frequency identification (RFID) tags to collect real-time core data including machinery entry/exit timestamps, operational zone positioning, and operator qualifications. It automatically generates digital ledgers synchronized with equipment certificates, inspection reports, and archival documentation. During equipment operation, multisource sensing terminals (vibration sensors, temperature monitors, HD cameras) dynamically monitor critical parameters of tower cranes, construction elevators, and other heavy machinery—including load capacity, structural integrity, and environmental risks—enabled by 5G networks for millisecond-level data transmission and edge computing.

Leveraging artificial intelligence (AI) recognition technology, the system performs real-time visual scanning of critical components (hook limiters, braking systems) through deep learning algorithms, automatically detecting structural deformations and latent defects. The predictive alert algorithm synthesizes multidimensional parameters (operational duration, maintenance cycles, environmental conditions) to build dynamic risk assessment models, issuing trilevel (red/yellow/blue) alerts to management personnel. When detecting excessive crane tilt angles or collision risks during multi-crane operations, the system immediately activates audiovisual alarms and terminates hazardous commands.

For hazardous construction equipment management, the system implements dual-prevention mechanisms for critical machinery. Automated shutdown protocols and

Serial Number	Equipment name	Entry time	Maximum lifting capacity
1	1# Tower base	2024.05	100t
2	2# Tower base	2024.05	100t
3	3# Tower base	2023.07	80t

Serial Number	Equipment name	Number of passengers	maximum payload
1	1# Construction Elevator	11 people	2200kg
2	2# Construction Elevator	11 people	2200kg
3	3# Construction Elevator	11 people	2200kg

Fig. 3 Mechanical management function diagram

emergency plans trigger when monitoring values exceed design thresholds. The maintenance module employs blockchain technology to immutably record lubrication processes, component replacements, and other servicing actions, creating tamper-proof digital logs. Managers access equipment health reports via mobile terminals, with AI-driven predictive maintenance recommendations improving service efficiency by over 60% compared to conventional methods while reducing equipment failures by 45%, ensuring intelligent safety governance throughout construction operations.

3.3 *Material Management*

As shown in Fig. 4, the material management module constructs a digital and intelligent management system, covering six core functions: material declaration statistics, type analysis, data statistics, incoming and outgoing rate monitoring, inventory dynamic tracking, and inventory difference analysis, to achieve precise control of material flow throughout the entire life cycle of engineering projects. The system deeply integrates IoT sensing devices and AI algorithms, connects with the material demand planning of various project departments' smart construction site platforms, synchronously integrates intelligent terminals such as weighing systems, RFID electronic tags, license plate recognition cameras, etc., collects in real time key data such as entry time, specification parameters, quality inspection reports, etc., of bulk materials such as concrete and steel, and automatically generates structured electronic ledgers. In the material acceptance process, optical character recognition (OCR) technology is used to automatically parse the delivery note information, completing compliance verification of model, quantity, and batch in a short period of time. Abnormal data triggers an immediate warning and freezes the signing process, preventing unqualified materials from entering the site from the source.

For material storage management, the system is based on UWB high-precision positioning technology to implement 3D digital modeling of key areas such as steel processing areas and template stacking yards. It combines AI video analysis to monitor the height of material stacking and the implementation of moisture-proof and fire prevention measures in real time. In terms of data analysis, the system constructs a multilevel intelligent model: a material demand prediction engine based on machine learning, which establishes turnover prediction models for building materials such as cement, sand, and gravel at different construction stages by mining historical engineering data, to assist in optimizing procurement cycles and inventory thresholds; and an inventory balance warning algorithm that automatically initiates cross project transfer recommendations when the difference in inventory of building materials of the same specifications between adjacent sections exceeds 25%.

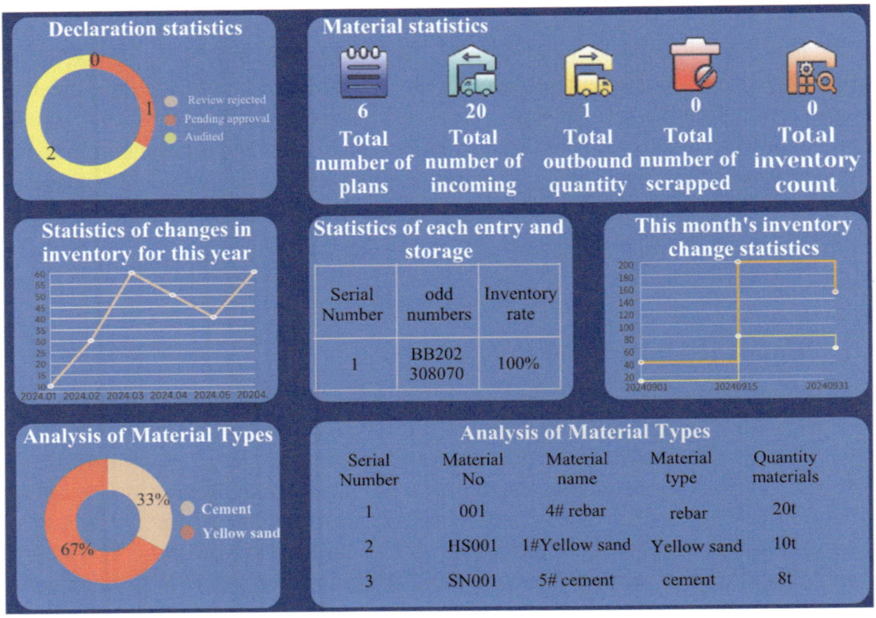

Fig. 4 Material management function diagram

3.4 Technical Management

As shown in Fig. 5, the technology management module, as a core component of engineering project management, encompasses secondary functional systems including plan management, Portable Document Format (PDF) drawings, computer-aided design (CAD) drawings, drawing management, drawing review, design changes, and technical inspections, establishing a closed loop digital control system spanning the entire project life cycle. Within the plan management subsystem, the system enables online compilation, multilevel approval, and version tracing of construction organization designs and specialized plans, while supporting remote collaborative review and electronic signatures for critical technical solutions by headquarters to ensure compliance and executability. For drawing management scenarios, the system integrates PDF and CAD dual engines to deliver innovative features such as intelligent version comparison, online annotation collaboration, and 3D visual inspection. Structured classification storage facilitates rapid cross-project drawing retrieval and reuse, effectively addressing traditional pain points of document fragmentation and delayed updates.

In process control, the drawing review module implements electronic approval workflows that enable multiparty collaborative review, automatically generating audit reports with real-time modification records. The design change management subsystem employs standardized forms to drive closed loop processes encompassing

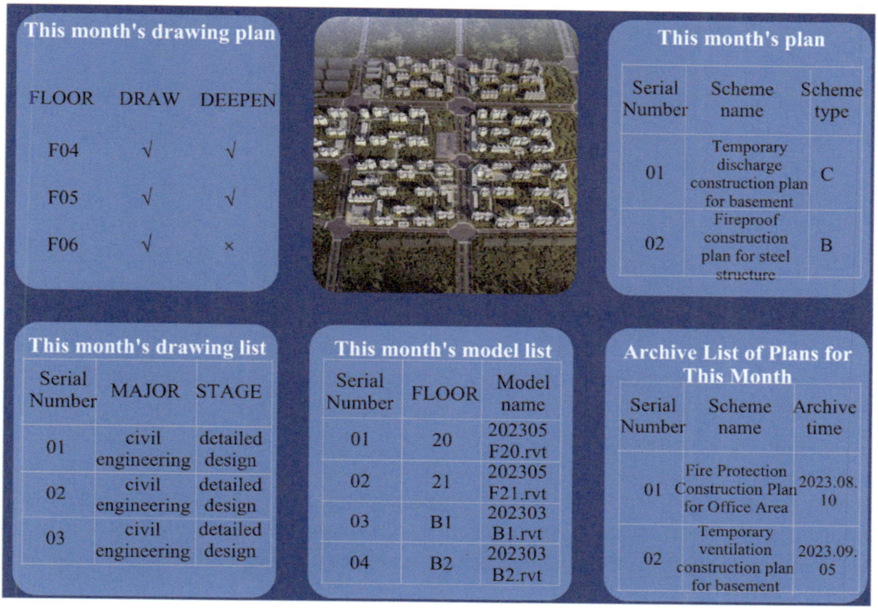

Fig. 5 Technical management function diagram

change requests, impact assessments, and approval execution, with technical parameters and cost data linked before and after modifications to ensure traceability. The technical inspection module innovatively integrates mobile GIS positioning and AI image recognition, allowing real-time photo documentation, automated issue categorization, and rectification tracking. Big data analytics generate periodic inspection reports to inform technical decision-making.

By eliminating data silos between project teams and headquarters, the system constructs a comprehensive management platform covering construction organization designs, drawings, measurement instruments, and testing protocols. The command layer gains real-time monitoring of technical key performance indicators (KPIs) across projects, employing multidimensional data dashboards to analyze regional risk trends and dynamically allocate expert resources. At the project level, standardized workflows and knowledge repositories significantly enhance technical disclosure efficiency and construction plan implementation quality. The platform automatically generates industry-compliant completion documentation archives, providing complete data assets for project acceptance and post-evaluation. This digital transformation empowers enterprises to achieve over 40% efficiency gains in technical management and 25% reduction in quality incidents, laying a solid foundation for intelligent construction strategies.

3.5 Environmental Management

As shown in Fig. 6, the system integrates IoT monitoring data to comprehensively analyze and manage on-site green construction practices by tracking key environmental parameters such as dust emissions, temperature/humidity levels, energy consumption, and water usage. Through real-time data collection from distributed sensors, it automatically identifies abnormal dust concentrations exceeding national standards, optimizes equipment operation schedules based on energy consumption patterns, and monitors water resource utilization efficiency. The platform dynamically adjusts construction strategies using intelligent algorithms, such as activating spray dust suppression systems when PM2.5 exceeds thresholds and scheduling high-energy processes during off-peak electricity periods. A dedicated carbon emission management module quantifies CO_2 equivalents from fuel combustion, material transportation, and machinery operations using internationally recognized calculation models. This module features real-time carbon footprint tracking, historical data comparisons, and automated generation of emission reduction suggestions, supporting the creation of carbon neutrality road maps. The system generates multidimensional green construction assessment reports and visual dashboards, enabling project managers to implement closed loop management of sustainable practices while providing regulatory authorities with verifiable low-carbon construction data to support environmental, social, and governance (ESG) compliance and green building certifications.

3.6 Quality Management

As shown in Fig. 7, the quality control module establishes a digital management system covering the entire project life cycle, comprising submodules such as Quality Data Overview, Smart Inspection Management, Standard Repository, and Technical Disclosure Management. The system automatically collects real-time on-site measurement data through IoT devices and integrates manually recorded defect information from inspections. Within the Quality Data Overview module, it generates dynamically updated multidimensional quality profiles that visualize key metrics across project sections, including qualification rate trends, typical issue distributions, and rectification closure rates.

Powered by AI algorithms, the system performs in-depth analysis of historical quality data to identify potential risks such as abnormal dispersion in concrete strength or fluctuations in rebar spacing compliance rates, enabling early warnings of quality deterioration trends. The Standard Repository consolidates national/industry technical specifications and intelligently delivers phase-specific acceptance criteria during construction.

For cross-project management, the system employs a quality performance evaluation model that benchmarks projects across eight dimensions, including process

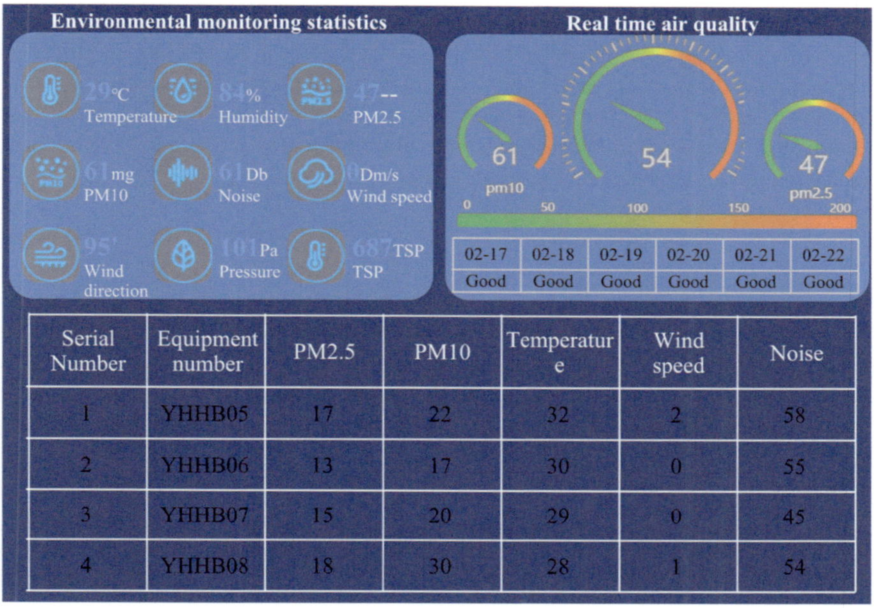

Fig. 6 Environmental management function diagram

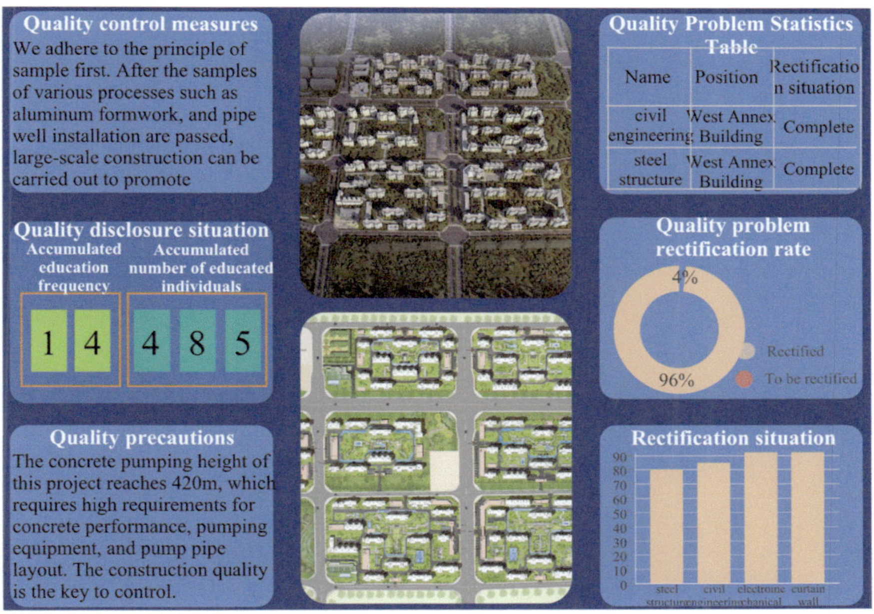

Fig. 7 Quality management function diagram

compliance rates, first-time acceptance rates, and quality-related costs, automatically generating quality performance rankings (red/black lists). In closed loop rectification management, the built-in BIM model automatically links quality defect locations. Through a mobile app, it enables full-process digital control encompassing task assignment, image-based process tracking, and multiparty acceptance verification, ensuring a defect closure rate exceeding 98%.

3.7 *Safety Management*

As shown in Fig. 8, the safety management module establishes an intelligent prevention and control system covering the entire construction process, integrating core functionalities such as safety status overview, smart inspection management, standardized safety briefings, safety knowledge base, dynamic risk source control, and early warning notifications.

The system collects real-time safety data across five categories, including equipment status and environmental parameters, through IoT devices such as stress sensors deployed across construction sites. Within the safety overview module, dynamic risk heatmaps visualize critical metrics such as hazard density across work zones, occurrence rates of three types of safety violations (unsafe commands, improper operations, labor discipline breaches), and integrity rates of protective facilities. For risk

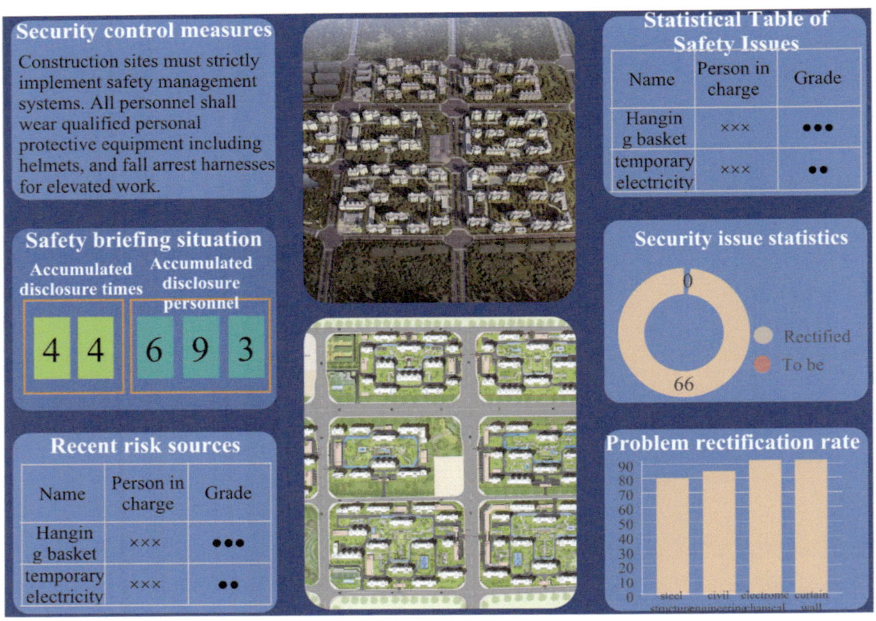

Fig. 8 Safety management function diagram

source management, the subsystem implements a three-tiered control system (red: 24-h monitoring; orange: daily inspection; yellow: weekly check) for major hazards like deep excavations and high-formwork supports. Targeted alerts are delivered to relevant work crews via an intelligent broadcast system.

In terms of education and training, the module features a smart question bank containing over 5000 test items, providing personalized learning materials tailored to different roles. For cross-project management, the system develops a Safety Performance Index (SPI) based on six dimensions, including hazard rectification rates and recordable injury rates per million man-hours, enabling quantitative benchmarking of safety management levels across projects.

3.8 Progress Management

As shown in Fig. 9, the Schedule Management module establishes a digital control system spanning the entire project life cycle, encompassing six core functionalities: panoramic schedule dashboard, intelligent work breakdown structure (WBS) decomposition, 4D schedule simulation, collaborative task dispatch, intelligent deviation diagnostics, and automated construction log generation. The system features a building information modeling–enterprise resource planning (BIM-ERP) bidirectional data interface that enables deep integration between Gantt charts created in Project software and BIM component libraries.

An IoT subsystem integrates smart weighbridges, GPS tracking, and equipment operation sensors to collect real-time progress metrics including concrete pouring volumes, delivery rate of transport vehicles, and pile foundation construction efficiency. After data cleansing through edge computing nodes, the schedule dashboard dynamically displays baseline vs. actual progress curves. When detecting rebar binding efficiency falling 15% below planned values, the system automatically triggers alerts and links to BIM models for root cause analysis. Project stakeholders can monitor milestone completion status in real time via mobile app, annotate responsibility attribution for delays online, and use blockchain technology to immutably record acceptance confirmations. The built-in AI diagnostics develop schedule risk prediction models to preemptively identify potential risks such as labor shortages' impact levels on main structural works. Monthly autogenerated multidimensional schedule analysis reports evaluate project management maturity through metrics like schedule compliance rate and resource balance index. These provide project owners with EPC contractor evaluation benchmarks, driving the transformation of schedule management from experience-driven to data-driven methodologies.

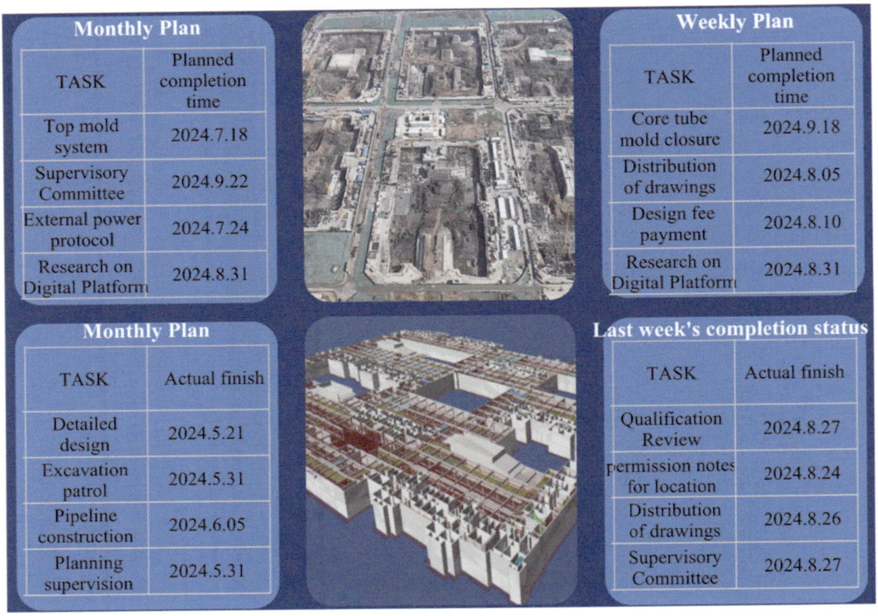

Fig. 9 Diagram of progress management function

4 Key Technologies

(1) Lightweight algorithm and comprehensive integration, transmission, and scheduling technology of multiprotocol interfaces for "edge end" multisource heterogeneous data:

① Data Compression and Preprocessing

Adopting layered compression technology: optimizes large 3D models to reduce file size while maintaining key details, segments and compresses sensor data (such as vibration and temperature), and retains core fluctuation characteristics through intelligent screening.

Developing an embedded data processing module: integration into intelligent gateway devices on construction sites, real-time filtering of abnormal data, and extraction of key indicators (such as device vibration peak), reducing data volume and cloud storage pressure.

② Multiprotocol Compatibility and Transmission Optimization

Building a protocol conversion system: supports mainstream industrial equipment communication protocols (such as Modbus, TCP/IP, etc.), automatically identifies device types, and unifies data formats. Automatically converts sensor data from different manufacturers into standard formats for cloud analysis.

Designing an intelligent transmission strategy: dynamically allocates network resources according to the requirements of construction scenarios and prioritizes the transmission of critical data (such as safety hazard alarms) [10].

③ Edge-Side Autonomous Operation Capability

Developing an emergency mode for network disconnection: When the network is interrupted, edge devices automatically enable local storage and perform basic analysis based on historical data to ensure uninterrupted basic monitoring.

Deploying lightweight self-checking function: Gateway devices automatically diagnose hardware status daily and notify operation and maintenance personnel through SMS after discovering faults, reducing the average fault response time to 2 h.

(2) Multi-object feature analysis technology for complex construction scenes based on computer vision:

In response to the problems of dense targets and multiple environmental interferences on construction sites, this study constructs a "perception analysis decision" full-chain visual intelligence system:

① 24/7 Environmental Adaptability

Adopting a multi-camera collaborative solution: combining ordinary cameras with infrared devices to solve the problem of blurry night construction monitoring.

Developing a light-adaptive technology: automatically adjusts image contrast and brightness and eliminates strong light reflection interference.

② 3D Visualization Progress Management

Building a dynamic progress comparison platform: overlays the construction progress with the 3D construction model and automatically labels the lagging areas of concrete pouring.

Establishing a database of violations: associates multiple security regulations, automatically identifies violations such as not wearing seat belts and lack of edge protection, and pushes rectification notices to the responsible person's mobile phone.

5 Pilot Project Case

The Yangzhen Shantytown Redevelopment Project covers an area of approximately 310.69 hectares, involving the relocation of 12,700 residents and the construction of 704,300 square meters of resettlement housing on a 35.03-hectare construction site. The entire project has a construction period of about 4 years. The general contracting consortium includes four companies within the China State Construction Engineering Corporation (CSCEC) system and two companies outside the system. With seven plots under simultaneous construction and extensive overlapping operations in a limited space, traditional project management methods proved challenging and high risk, making the adoption of digital tools for project management particularly urgent.

By implementing a BIM + Smart Construction Site platform, the project enhanced digital and intelligent engineering control, providing critical support for successful delivery. The use of a digital command and decision-making platform for construction projects enabled the client to achieve unified management through a "single-screen overview," addressing oversight of safety, quality, progress, and other key metrics, as well as management of supervisors, contractors, and other stakeholders. As a practical tool for large-scale project management, it significantly improved governance capabilities and pioneered innovative management models for complex projects, offering valuable insights for future endeavors.

6 Conclusion

Under the overarching trend of industry digital transformation, the rapid development of BIM, IoT, and AI technologies has established a robust technical foundation for building smart construction sites. By developing a building information modeling + Artificial Intelligence of Things (BIM + AIoT)-based smart construction platform, the current overreliance on manual management can be reversed, significantly enhancing the refinement, digitization, and intelligence of construction sites, improving project safety standards and advancing the delivery of high-quality engineering projects. Looking forward, as BIM technology and new-generation information technologies continue to evolve, BIM + AIoT-based smart construction platforms are poised to be widely adopted across more engineering and construction projects. This will propel the industry toward a high-quality development trajectory.

References

1. Dong, Y.: Current status and solutions for overcapacity in construction industry. China Econ. Trade Herald **23**, 18–19 (2016)
2. Feng, W.: Application of fine management in construction engineering. Theor. Res. Urban Constr. **29**, 59–61 (2024)
3. Xu, K., Chen, X.: A "Five-High" risk control system for major accident prevention. J. Wuhan Univ. Technol. (Inf. Manage. Eng.) **39**(6), 5 (2017)
4. Sun, Y.: Analysis and prevention of construction safety accidents. China Constr. Metal Struct. **04**, 52–53 (2021)
5. Zheng, Y., Yi, C., Liu, M., et al.: Impact of prior knowledge and noise on hazard identification in construction. Safety **45**(11), 68–73 (2024)
6. Li, J.: Innovation in project management based on big data. Commod. Qual. **36**(020), 35–38 (2015)
7. Xie, Y.: Research and application of a Beidou+BIM-based smart construction site platform. Urban Geotech. Investig. Surv. **6**, 55–58+63 (2024)

8. Huang, Z., Zheng, H., Zhong, X.: Application of 5G+ smart construction site BIM cloud platform in project management. Intell. Build. Smart City **11**, 95–97 (2024)
9. Guo, Z.: Technical analysis of smart construction site based on BIM+GIS. Resid. Ind. **08**, 23–25 (2023)
10. Zhang, G., Xiao, J., Li, Y., et al.: Research and design of multi-protocol edge computing gateway. Manuf. Autom. **46**(12), 163–170 (2024)

Design and Implementation of the Low-Altitude Intelligent Traffic Management System for Urban Environments

Xiaoxia Zhou, Tao Tang, Xiang Zou, Xi Fu, and Xinyi Yang

Abstract Facing challenges in urban low-altitude traffic regulation such as decentralized operators, disordered flight activities, and frequent airspace conflicts, this study designs and implements a low-altitude intelligent traffic management system (LITMS) for urban environments. The system adopts a "user side + management side" dual-end collaborative architecture: The user side enables whole-process electronic management including aircraft/operator registration and flight plan submission through a one-stop service portal; the management side builds a city-level digital twin platform based on multisource heterogeneous data (including unmanned aircraft dynamic data, three-dimensional [3D] real-scene models, laser point clouds, and meteorological information) to achieve large-scale flight dynamic monitoring and alert warnings. In practical deployment in Chengdu, the system successfully completed real-time monitoring of over 2 million unmanned aircraft flights in 2024, verifying its reliability and practicality in complex urban environments. The research results provide an effective technical solution and implementation example for urban low-altitude intelligent traffic management, which is of great significance for promoting the standardized development of urban air traffic.

Keywords Low-altitude · Unmanned aircraft · Control system · Spatial grid · Real-time monitoring

1 Introduction

Low-altitude economy refers to a comprehensive economic form that takes the low-altitude flight activities of manned and unmanned aircraft (unmanned aircrafts) in low-altitude airspace as the traction and drives the integrated development of related fields. The operating environment for low-altitude flight activities in urban areas is

X. Zhou · T. Tang (✉) · X. Zou · X. Fu · X. Yang
Sichuan Provincial Key Laboratory of Low-Altitude Traffic Intelligent Control, The Second Research Institute of CAAC, Chengdu, China
e-mail: tangtao@caacsri.com

K. Subramaniyam and M. Paprzycki (eds.), *Smart Cities and the Internet of Things*, Internet of Things, https://doi.org/10.1007/978-3-032-08290-9_20

complex, with urban building canyons, low-altitude electromagnetic interference, complex low-altitude meteorological conditions, and dense populations all being important factors affecting low-altitude flight safety. According to statistics from the Civil Aviation Administration of China, by the end of 2023, the industry had registered a total of 1.267 million unmanned aircrafts, with annual cumulative flight hours reaching 23.11 million hours [1]. With the rapid development of application scenarios such as urban air mobility (UAM), unmanned aircraft logistics distribution, emergency rescue, and urban governance, low-altitude flight activities are gradually showing characteristics of large flow, strong coupling, and high time variability, posing severe challenges to the existing low-altitude traffic management system. The International Civil Aviation Organization (ICAO) emphasized in the Implementation Guidelines for Unmanned Aircraft System Traffic Management (UTM) that low-altitude traffic management in urban environments needs to break through the three major technical bottlenecks of "difficult data fusion, low response timeliness, and narrow regulatory coverage" [2]. Therefore, to ensure smooth, efficient, safe, and reliable operation, the traditional "human-managed" model is unsustainable, and it is necessary to develop digital, autonomous, and collaborative low-altitude intelligent traffic control modes and methods.

In terms of low-altitude intelligent traffic control technology and system construction, scholars at home and abroad have carried out a series of exploratory work. Current international research on low-altitude intelligent traffic management platforms emphasizes collaborative frameworks, artificial intelligence (AI) integration, and regulatory harmonization. The US UTM system focuses on scalable unmanned aircraft operations in segregated airspace, leveraging AI for real-time deconfliction and beyond visual line of sight (BVLOS) coordination [3]. The European Union (EU)'s U-Space framework integrates advanced services like dynamic airspace allocation and conflict resolution [4] with projects like U-space European COMmon dEpLoyment (U-ELCOME), demonstrating multi-unmanned aircraft coordination in urban environments [5]. Singapore's Traffic Management of Unmanned Aircraft System (TM-UAS) initiative employs swarm intelligence and blockchain for risk assessment [6], while South Korea's Seoul Metropolitan Government utilizes AI-powered unmanned aircrafts and 4G networks for real-time traffic monitoring [7]. Australia's CAELUS Project integrates cellular infrastructure for reliable unmanned aircraft communication [8]. In China, a series of control technology breakthroughs have been achieved in urban unmanned aircraft safety separation [9], flight plan coordination [10], and path planning [11, 12]. In terms of system development, the independently developed National Unmanned Aircraft Operation Management Platform (UOM) can realize dynamic monitoring of cooperative unmanned aircrafts nationwide [13]. To promote the healthy and orderly development of the unmanned aircraft industry and safeguard aviation security, public safety, and national security, the State Council and the Central Military Commission jointly issued the Interim Regulations on the Flight Management of Unmanned Aircraft in May 2023 [14]. These regulations establish a national legal framework for standardizing unmanned aircraft operations across four dimensions: personnel (operators), equipment (airworthiness of aircraft), environment (airspace), and management (flight activity oversight). In

December 2023, the National Airspace Basic Classification Method [15] was promulgated and implemented. Aligned with the International Civil Aviation Organization's airspace classification standards, it restructures China's national airspace system by designating Class G and Class W airspace to empower low-altitude drone operations.

At present, domestic and foreign system designs do not fully consider the complex urban environmental factors and cannot achieve refined control and diversified services for urban low-altitude airspace, making it difficult to support the large-scale and clustered operation of urban low-altitude aircraft in the future. Therefore, the design and implementation of low-altitude intelligent traffic control systems in urban environments are of great significance for promoting the development of the low-altitude industry, and the market prospects are very broad.

2 Demand Analysis

2.1 Business Requirements

From the perspective of industrial structure, the low-altitude economy can be divided into four major components: low-altitude aircraft manufacturing, low-altitude flight service support, low-altitude infrastructure, and low-altitude operational services. Among these, low-altitude flight service support serves as the hub connecting other components and forms the prerequisite for the safe and orderly development of the low-altitude economy.

The low-altitude intelligent traffic management system will provide urban unmanned aircraft operators with high-quality low-altitude flight service support, including regulatory services, traffic services, and commercial services. Through intelligent monitoring and other means, it will assist industry regulatory authorities in safety supervision and help establish a secure and orderly low-altitude flight environment, thereby promoting the vigorous development of the low-altitude economy industrial ecosystem.

2.2 Information Exchange and Sharing Requirements

Based on this system, cross-regional, cross-hierarchical, and cross-departmental data exchange will be achieved with civil/military airspace control departments, industry regulatory authorities, and local administrative departments. On one hand, data sharing with various departments will be realized; on the other hand, the system will act as a data exchange hub to facilitate interoperability with multiple platforms and systems, supporting the safe operation of unmanned aircraft. Key interfacing platforms include the following:

- Integration with civil/military airspace control departments to enable unified flight plan submissions, access to controlled/suitable airspace, flight activity authorization confirmations, and transmission of temporary restricted airspace information for unmanned aircraft operators.
- Data interconnection with the National Unmanned Aircraft Integrated Supervision and Service Platform (Civil Unmanned Aircraft Comprehensive Management Platform) to obtain real-time flight dynamic data of unmanned aircrafts within jurisdictional areas.
- Interface with industry regulatory authorities to acquire event handling records, alerts, and emergency notifications.
- Connection with meteorological platforms to integrate various types of aviation weather data.
- While implementing diverse information exchanges and sharing, coordinate and allocate responsibilities for various types of information generated during low-altitude flight activities to avoid redundant data infrastructure across platforms. Data integration will be performed to create sustainable data products that support modernization and informatization requirements for low-altitude flight management.

2.3 Performance Requirement

- Number of simultaneously processed and displayed unmanned aircraft flight monitoring data $\geq$ 10,000 units (scalable based on market demand).
- Average response time for user access to functional points $\leq$ 1 s.
- Average user login platform response time $\leq$ 1 s.
- Average response time for simple operations (query/add/delete) $\leq$ 1 s.
- Average response time for complex operations $\leq$ 2 s.
- Average response time for complex operations $\leq$ 1 s.
- System continuous uptime $\geq 7 \times 24$ h.

3 System Design

3.1 Overall Architecture

Based on the urban low-altitude data foundation, we construct a low-altitude intelligent traffic management system for urban environments with "one data foundation and three operational subsystems" as its core. The three operational subsystems are the low-altitude traffic flight service subsystem, the low-altitude traffic supervision subsystem, and the low-altitude public safety management subsystem. Among them, the low-altitude traffic flight service subsystem will be equipped with web-based and mobile declaration services. The overall architecture diagram is shown in Fig. 1.

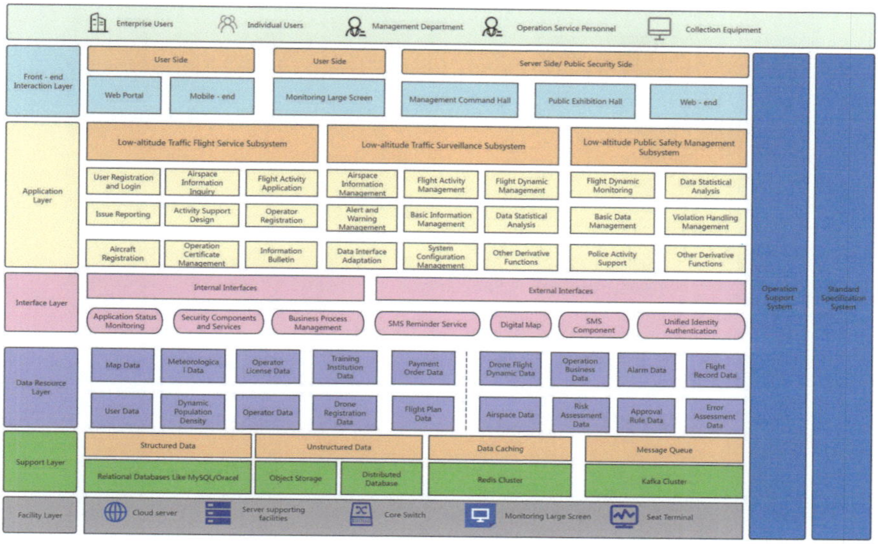

Fig. 1 System's overall architecture diagram

1. The low-altitude traffic flight service subsystem is built to provide low-altitude users with services such as airspace route applications, flight plan declarations, communication and navigation, information, and meteorological services.
2. The low-altitude traffic supervision subsystem is constructed to achieve functions including access approval and process supervision. It will interconnect with the flight service platform and higher-level supervision platforms to realize dynamic transmission of information data and strengthen whole-process supervision of operations.
3. The low-altitude public safety management subsystem is established to support the implementation of urban air traffic safety management measures and ensure the safe operation of urban low-altitude traffic.

3.2 Technical Architecture

The system fully leverages the user's existing foundational platform and adopts a microservice architecture. Each business application operates as an independent microservice, with redundant microservice configurations implemented for core business functions. The entire system consists of multiple microservices. For microservice technology, Spring Cloud is used as the development framework. Application programming interfaces (APIs) or message middleware serves as the means of data communication between subsystems. A service gateway functions as the backend service interface for the system, enabling service routing selection, while

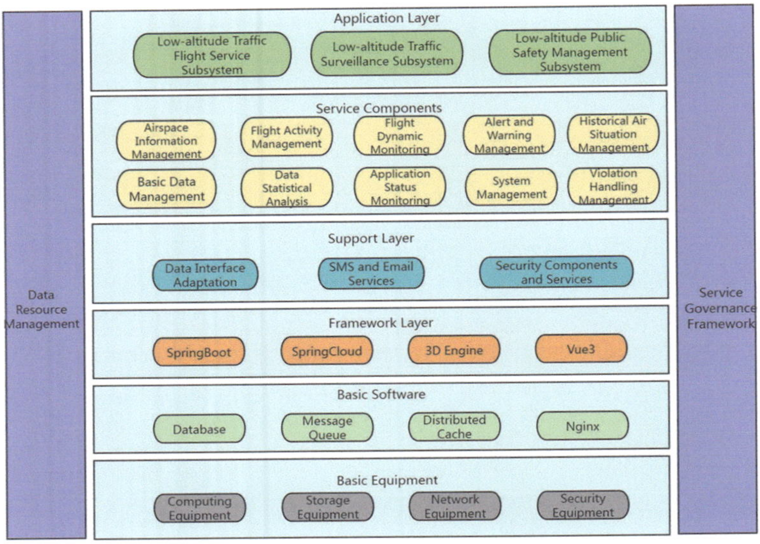

Fig. 2 System's technical architecture diagram

load balancing is achieved through nginx reverse proxy. In alignment with business requirements and based on existing industry standards and system integration specifications, the system processes data related to database connection information, data file exchange protocols, authentication methods, encryption/decryption methods, and data interface protocols. It collects business application-related data and interfaces with multiple internal and external systems. The system classifies and processes source system data in terms of scope, structure, business meaning, business rules, system relationships, data structure, data dictionaries, and collection methods. It designs extraction strategies for source system data and implements unified management. The system technical architecture diagram is shown in Fig. 2.

3.3 Network Architecture

The system deployment adopts a core switching and access switching—two-layer switching architecture—with separate core switching zone, Internet access zone, government network access zone, and private network access zone. The core switching zone connects to the government network switching zone and private network switching zone through cross-network secure exchange and establishes project security domains through secondary switching subnets. The system network architecture diagram is shown in Fig. 3.

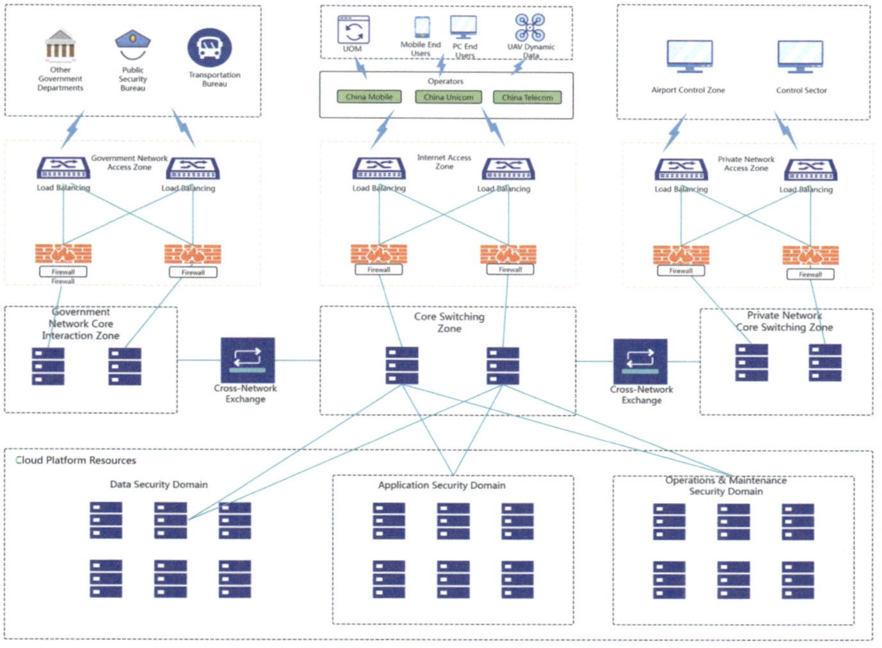

Fig. 3 System's network architecture diagram

3.4 *Functional Design*

The system is built upon the urban low-altitude data infrastructure, establishing a low-altitude intelligent traffic management system centered around "one data infrastructure and three operational subsystems." The system will reserve data interfaces with various external platforms to facilitate information exchange and collaborative operations between upper- and lower-level systems. It will also maintain interfaces for integrating external data sources, continuously enriching the platform's data types to provide robust data support for regulatory services.

Low-Altitude Data Infrastructure

The data infrastructure of the low-altitude intelligent traffic management system serves as the foundational platform supporting system operations, enabling data interconnection with civil aviation authorities and various government departments. It establishes a multisource heterogeneous business data infrastructure, incorporating UOM-related data, detection equipment data, low-altitude meteorological data, geographic information system (GIS) map data, Automatic Dependent Surveillance–Broadcast (ADS-B) general aviation aircraft data, and real-time image data. This infrastructure provides data support for system operations, facilitates upper-layer data retrieval and visualization applications, enhances data analysis and application capabilities, and enables comprehensive monitoring of low-altitude aircraft, air traffic management, and early warning/alarm services for integrated traffic management.

Low-Altitude Traffic Flight Service Subsystem

Serving as the unified public portal for urban unmanned aircraft operations, the low-altitude traffic flight service subsystem enables operating enterprises to conduct unmanned aircraft flight-related services and information queries through both personal computer (PC) and mobile interfaces. Key functions include airspace information inquiry, flight activity applications (airspace demand submission/flight plan submission), operator management, aircraft registration, operation certificate management, notification, and announcement queries.

Low-Altitude Traffic Supervision Subsystem

Airspace Information Management: The airspace information includes permissible airspace, restricted airspace, temporary restricted airspace, key monitoring airspace, and flight routes. Among these, permissible airspace and restricted airspace are uniformly issued by the national air traffic management authority; temporary restricted airspace and flight routes are compiled and summarized by the platform regulatory department, approved through the traffic supervision subsystem, and finally published to the low-altitude traffic flight service subsystem; key restricted airspace can be manually delineated and entered through the traffic supervision subsystem, serving as a reference for airspace review.

Flight Dynamic Monitoring and Alerts: All regulatory service participants can monitor the real-time flight dynamics of unmanned aircraft within their respective jurisdictions through the platform. Based on each department's requirements, different alert rules can be configured to determine violations, and short message service (SMS) alerts can be sent for notification purposes.

Flight Activity Approval: Regulatory service departments shall verify and approve flight activity applications submitted by unmanned aircraft operators according to their respective responsibilities, ultimately ensuring lawful and compliant operations of unmanned aircraft within urban areas.

Basic Information Management: This module manages and maintains real-name registration data, operator qualification records, operator credit information, pilot credit information, and operation certificate details, providing critical reference for flight activity applications/approvals and violation handling.

Statistical Analysis Dashboard Display: Integrating multiple data domains including flight activity, flight volume, alert violations, airspace routes, and foundational data, this module performs queries and data fusion across thematic datasets to generate consolidated outputs for large-screen visualization.

System Management and Application Monitoring: The system management module includes functions such as permission management, role management, log management, file management, workflow management, system configuration, authentication and authorization, and system monitoring, providing robust support for all departments using the platform. Application monitoring primarily tracks basic information, service status, service response performance, hardware utilization, and fault conditions to ensure stable platform operation.

Low-Altitude Public Safety Management Subsystem

Low-altitude Public Safety Management Subsystem: The low-altitude public safety management subsystem is a dedicated subsystem for public security departments, visible only to relevant management accounts of public security authorities. Taking unmanned aircraft industry management as the starting point, based on unmanned aircraft-related information intelligence, and aiming at unmanned aircraft flight activity monitoring and disposal, it has comprehensive management, monitoring, analysis, and disposal functions for unmanned aircraft operating enterprises, flight activities, operators, and control equipment. It truly achieves comprehensive and efficient control of the three elements of unmanned aircraft–human aircraft–environment and responds to the increasing unmanned aircraft safety risks by integrating various technical means, improving public safety levels, and ensuring social order.

4 Core System Technologies

4.1 Flight Dynamic Data Fusion Processing

Real-time unmanned aircraft flight dynamic data fusion processing technology is one of the key technologies determining the success of low-altitude intelligent traffic management system construction. It must not only consider the distributed computing problems of high-concurrency flight dynamic data caused by the future increase in unmanned aircraft flight volume but also consider how to quickly perform spatial operations based on urban routes and airspace data when processing individual flight dynamic data to determine and generate alerts. The design of the data fusion processing algorithm itself is also crucial. Based on extensive experience in information system construction and flight data fusion processing, this study designs a highly available, high-performance, and scalable flight dynamic data fusion processing solution, as detailed in Fig. 4. Verified results show that it can achieve high-concurrency flight dynamic data fusion processing at a level of 50,000 per second, with processing time for individual flight dynamic data being less than 100 ms when performing spatial operations and alert determination based on millions of airspace data points.

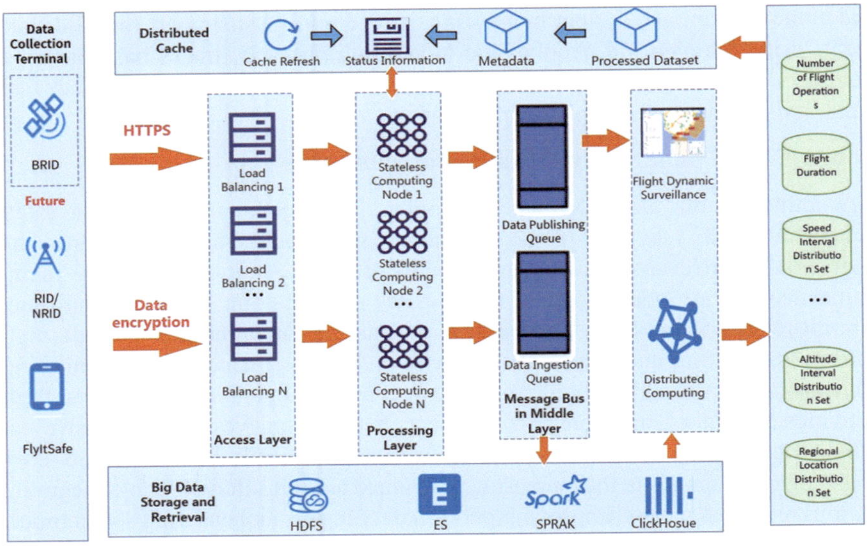

Fig. 4 Flight dynamic data fusion processing technology diagram

4.2 Spatial Grid-Based Digital Base Platform

With the development of the low-altitude (e.g., unmanned aircraft) industry, the number of low-altitude aircraft is experiencing explosive growth, urgently necessitating the establishment of an information system for effective management and control. Compared to traditional air traffic control systems for manned aircraft, which have capacity requirements on the order of 10^3, this system's capacity needs to be increased by at least 1–2 orders of magnitude. This places significant pressure on the retrieval and computation of spatial data based on latitude–longitude representations.

To address this, continuous space is partitioned into discrete grids, with each grid corresponding to a unique code. The positional relationships between airspaces are transformed into topological and directional relationships between grids. Complex distance calculations based on latitude–longitude coordinates are simplified into arithmetic operations (addition, subtraction, multiplication, and division) between grids, ultimately reducing to computations between codes.

Figure 5 illustrates the spatial grid-based airspace representation. Through this gridding method, the efficiency of large-scale unmanned aircraft flight conflict detection and airspace conflict detection can be improved by 1–2 orders of magnitude.

Fig. 5 Spatial grid-based airspace representation diagram

5 System Application

The Chengdu Low-altitude Traffic Management Service Platform, developed with the low-altitude intelligent traffic control system as its core, has completed its first operational test on June 20, 2024, marking it as the first city-level low-altitude digital governance system in China. Jointly established by Tianfu New Area and Chengdu Communications Investment Group, among others, the platform adopts a "1 + 3 + N" architecture and comprises seven core functional modules (including operational situation monitoring and flight dynamic surveillance), capable of real-time monitoring of aircraft movements within a 50-km range (see Fig. 6). During testing, it successfully facilitated China's first helicopter route interconnection between two airports (Tianfu/Shuangliu) and validated the urban-to-Qingcheng Mountain scenic area route.

Currently, the platform has achieved three key functionalities:

Daily handling of 300 + logistics delivery flights and integration of regulatory data from 28,000 unmanned aircrafts citywide.

Enabling a 15-min rapid response in emergency relief operations, such as deploying 278 unmanned aircraft sorties during the 2024 Luding earthquake rescue.

Supporting electric vertical takeoff and landing (eVTOL) commercial application testing, with AutoFlight completing full-tilt flight trials of the AE200 model, laying the foundation for future urban air mobility.

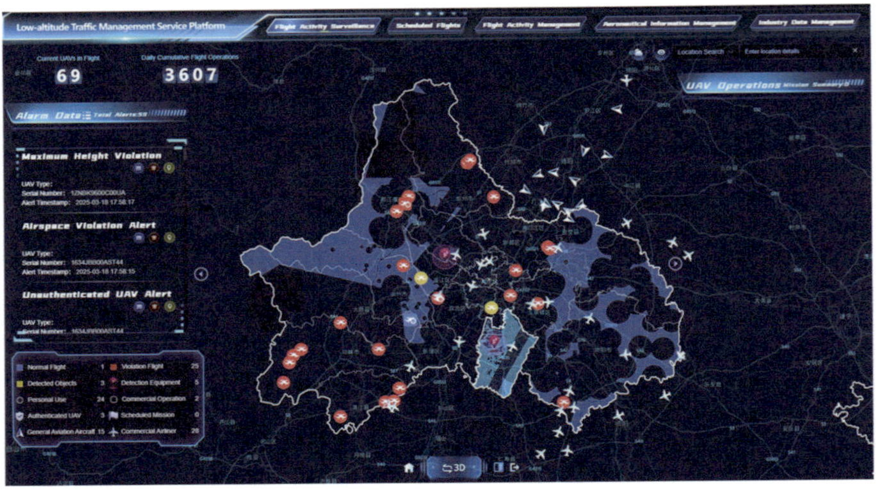

Fig. 6 System's application flight dynamic monitoring interface diagram

6 Conclusions

The low-altitude intelligent traffic management system (LITMS) designed and implemented in this study adopts a collaborative "user side + management side" architecture, integrating three core subsystems—low-altitude traffic flight service subsystem, low-altitude traffic supervision subsystem, and low-altitude public safety management subsystem—to establish a full-process digital management platform for urban low-altitude traffic. Leveraging multisource data fusion, spatial grid-based modeling, and high-concurrency processing technologies, the system effectively addresses core challenges such as fragmented data integration, slow regulatory response, and frequent airspace conflicts in complex urban environments. During its deployment in Chengdu in 2024, LITMS successfully monitored dynamic operations of over 2 million unmanned aircraft flights, supporting logistics delivery, emergency relief (e.g., a 15-min rapid response during the Luding earthquake), and eVTOL commercial trials, thereby validating its reliability and practicality in high-density, tightly coupled urban airspace. By constructing a digital twin foundation and cross-departmental data sharing mechanisms, LITMS provides a replicable technical paradigm for standardized and intelligent urban low-altitude traffic management.

Future research should focus on enhancing the system's capabilities in ultra-large-scale unmanned aircraft swarm coordination, cross-domain heterogeneous platform interoperability, and extreme environment adaptability. Integrating 5G/6G communications, quantum computing, and autonomous decision-making technologies could enable dynamic optimization of airspace resources and fully automated traffic flow scheduling. Deeper integration with smart city systems and urban air mobility (UAM) networks may facilitate air–ground–underground multimodal traffic integration. Additionally, establishing a globally interoperable standardized regulatory framework is critical to addressing challenges such as cross-border flights and emerging security threats, laying a technical foundation for the globalization of the low-altitude economy.

Acknowledgements This work was supported by Sichuan Science and Technology Program under Grant No. 2024ZDZX0044.

References

1. Civil Aviation Administration of China (CAAC): Statistical Bulletin on the Development of Civil Aviation Industry 2023. Beijing (2024)
2. International Civil Aviation Organization (ICAO): Manual on Remotely Piloted Aircraft Systems (RPAS). Montreal (2023)
3. NATS Services: Advancing Aviation Through Integration of Drones in the UK's Low-Level Airspace, White Paper (2025)
4. Lappas, V., et al.: EuroDRONE, a European UTM testbed for U-space. In: 2020 International Conference on Unmanned Aircraft Systems (ICUAS), pp. 1766–1774. Athens, Greece (2020). https://doi.org/10.1109/ICUAS48674.2020.9214020
5. Razali, M., et al.: Machine learning for traffic flow prediction. J. Big Data **12** (2025)
6. NTU Singapore: Traffic Management of Unmanned Aircraft Systems (TM-UAS), Technical Report (2025)
7. Seoul Metropolitan Government: AI-driven unmanned aircraft traffic monitoring system. Telecommun. J. **45** (2025)
8. Airservices Australia: Digital air traffic management for urban unmanned aircraft operations. J. Air Transp. Manage. **89** (2024)
9. Chen, Y., Zhang, J., Wang, Z., et al.: Safety separation calculation model for low-altitude multi-rotor unmand aircrafts based on collision avoidance maneuver strategy. Acta Aeronaut. Astronaut. Sin. 1–21 (2025). http://kns.cnki.net/kcms/detail/11.1929.V.20250425.1241.004.html
10. Hua, X., Site, H., Jianan, Y., et al.: Collaborative simulation and optimal allocation method for urban low-altitude unmand aircraft flight plans. Acta Aeronaut. Astronaut. Sin. **45**(19), 269–291 (2024)
11. Zhang, H., Li, H., Liu, H., et al.: Route planning for logistics unmand aircrafts in urban areas. J. Transp. Syst. Eng. Inf. Technol. **20**(06), 22–29 (2020). https://doi.org/10.16097/j.cnki.1009-6744.2020.06.003
12. Zhao, Y., Gu, R., Ren, X.: Method for small unmanned aerial vehicles path planning considering urban low-altitude wind fields. J. Beijing Univ. Aeronaut. Astronaut. https://doi.org/10.13700/j.bh.1001-5965.2024.0281
13. Jia, Y.: Preliminary exploration of traffic management scheme for unmanned systems in low altitude airspace. J. Aeronaut. 1–25 (2025). http://kns.cnki.net/kcms/detail/11.1929.v.20250428.1511.006.html
14. The Central Military Commission of the State Council of the People's Republic of China: Interim Regulations on the Flight Management of Unmanned Aircraft. National Order No. 761, May 2023
15. Civil Aviation Administration of China (CAAC): National Air-Space Fundamental Classification Method (2023)

Design of Underground Pipe Gallery Monitoring and Early Warning System Based on Digital Twin Technology

Yingxia Li and Weiying Hao

Abstract As a key component of urban infrastructure, urban underground pipe galleries are related to people's livelihood and well-being. There are problems such as difficult planning, construction, monitoring, and maintenance. Based on digital twin technology, this article mainly uses the Lora communication protocol to achieve underground blind spot data transmission, cloud-based data controllable strategies to achieve automatic monitoring and early warning, and automatic real-time push of work orders. It creates a comprehensive smart pipe gallery with "one gallery, multiple warehouses, and one supervision," providing theoretical and practical references for the design and implementation of smart pipe galleries.

Keywords Intelligent pipe gallery · Digital twin · Monitoring · Warning

1 Introduction

The intelligent pipe gallery system integrates advanced technologies such as the Internet of Things, modern sensing technology, cloud computing, computer communication technology, virtual reality, and artificial intelligence, achieving real-time monitoring, operation management, and security warning functions for the internal pipelines, environment, equipment, and operating status of the comprehensive pipe gallery. Currently, the system is showing a rapid development trend in both domestic and international markets.

With the continuous promotion of national digital city construction, stricter requirements have been put forward for the digital construction and management of urban public facilities. Given the unique distribution environment of underground pipe galleries, there are many challenges in their planning, construction, monitoring,

Y. Li
Hubei Urban Construction Vocational and Technical College, Wuhan, China

W. Hao (✉)
Hubei University of Technology, Wuhan, China
e-mail: 782336498@qq.com

K. Subramaniyam and M. Paprzycki (eds.), *Smart Cities and the Internet of Things*,
Internet of Things, https://doi.org/10.1007/978-3-032-08290-9_21

warning, and maintenance. The frequent occurrence of accidents caused by natural disasters and human factors has made it an urgent research topic on how to efficiently manage, monitor, and warn, as well as how to use effective information for daily, emergency, and wartime planning, deployment, monitoring, warning, and emergency response. The research in this field not only has important theoretical and practical significance but also Heralds enormous market potential.

2 Related Studies

Based on digital twin multidimensional models and twin data, core application services that meet the safety management of urban underground comprehensive pipe galleries are provided, thereby achieving twin empowerment and improving quality and efficiency [1]. This study points out that the use of IoT technology can improve construction efficiency and promote the modernization of pipe gallery systems [2]. The study uses qt + vs2010 + SQLite and Unity3D software to create a system, a dual compartment underground pipe gallery model is established, and intelligent query function is implemented [3]. By comparing the Code for Construction, Detection and Acceptance of Building Intelligent Engineering (DBJT 15-147—2018) and the Code for Quality Inspection of Urban Rail Transit Intelligent System Engineering (DBJ/T 15-196—2020) in Guangdong Province, and combining with the engineering practice of underground pipe gallery, the study summarizes the main intelligent systems of underground pipe gallery and constructs a set of current provincial standards to guide the detection work [4]. The results indicate that the mobile three-dimensional (3D) laser scanner can significantly improve operational efficiency and meet measurement accuracy requirements in underground pipe gallery measurement [5]. The study found that the application of 3D visualization technology can greatly improve the efficiency and accuracy of underground pipeline detection [6]. References [7–9] provides a detailed analysis and discussion of the development, application, and challenges of digital twin technology, services, platforms, and service system design frameworks. It offers corresponding solutions. References [10–12] presents its own insights on the application of digital twins in smart city management. X provides a framework for intelligent city digital twins and reflects on future trends. This study offers methods and technical references for the construction of data platforms.

Based on relevant literature and domestic and foreign technological research, this study uses sensing technology, cloud computing, computer communication technology, virtual reality, and other technologies to achieve real-time monitoring, operation management, and safety warning of pipelines, environment, equipment, and operation in urban underground pipelines and to carry out unified daily monitoring and warning management. Digital twin technology is used to build a visual operation and maintenance platform for real-time unified classification and management of various equipment in pipelines, capturing abnormal situations in a timely and controllable manner, and handling emergency situations.

3 System Architecture

3.1 Overall Objectives of the Work

The integrated application of virtual reality, IoT, 5G, and cloud platform technologies enables the establishment of a unified data platform and digital twin display interface. This platform provides standardized data interfaces to maximize the elimination of information silos, achieves real-time automated detection, early warning, and emergency response capabilities, while featuring excellent scalability and maintainability.

The main goals are as follows:

- Cloud platform with a unified data interface.
- A digital twin platform for unified data display and management.
- Internship according to the efficient underground data collection and transmission functions.
- Realize the real-time automatic detection and early warning function.
- To achieve the precise positioning and query function.
- Realize the function of self-construction single push.
- Good data program separation, scalable, and easy maintenance.

In the process of achieving goals, the following four principles are mainly followed:

- Intelligent: Automatically detect anomalies and achieve rapid response and disposal.
- Data: Data analysis and prediction to improve the accuracy and efficiency of response.
- Visualization: Facilitates real-time monitoring and command of emergency personnel.
- Safety: Advance technical intervention to ensure the safe operation of the equipment.

3.2 Technical Architecture of the Works

The system supports the public cloud or the private cloud, which includes the operating system of the cloud service platform, the database management system, the data storage, and other devices. Cloud deployment is divided into two components: cloud-based and on-premises, with the latter handling collection services. The acquisition server is connected with the management host of each intelligent subsystem through the internal local area network (LAN) and is connected with the cloud through the private network or public network, using browser/server (B/S) architecture. The system's technical architecture diagram is shown in Fig. 1.

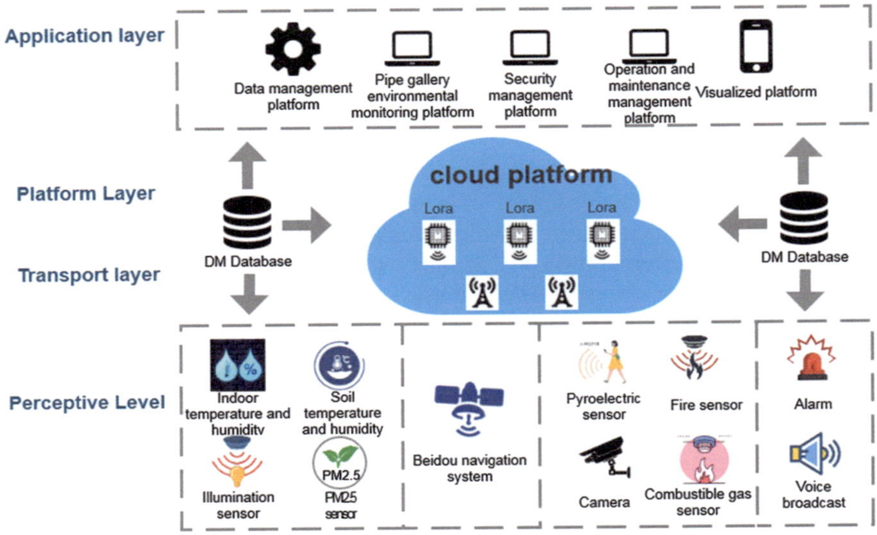

Fig. 1 The system's technical architecture diagram

The system achieves seamless integration of all modules through an ESB data bus or middleware, meeting all requirements for 3D visual operations, data integration, and intelligent management. The digital twin platform unifies data display and delivers comprehensive data analysis and reporting functions, covering sensitive equipment parameters, environmental metrics, operational status, and geographic information. By applying multiple analytical methods, it generates actionable insights that support decision-making at all management levels and enables full monitoring, early warning, and emergency response capabilities. The key technologies for each function are shown in Table 1.

The data acquisition layer adopts wired and wireless communication protocol services, provides industrial-level real-time database, and supports the efficient access of upper-level applications. According to the characteristics of underground wireless signal transmission, LoRa communication protocol is adopted [13, 14].

Table 1 The key technologies for each function

Function	Key technology
Data acquisition	Sensor technology, radio frequency identification (RFID) technology, IoT technology, geographic information system (GIS)
Data transmission	LoRa, NB-IoT, Bus, WiFi,5G
Data storage	Cloud platform, big data technology
Data application	Unity, 3D visualization technology, cloud platform, data visualization technology, 3D light detection and ranging (LiDAR) scanning modeling

Through the service layer, the system's functionality is expanded to enable multi-format data collection and sharing, cross-subsystem data exchange, and external interface services including voice service modules, SMS service modules, and global event processing modules. These server-hosted services directly correspond to and support the various functional implementations at the application layer.

The system fully supports cloud computing, enabling users to access the required applications with various terminals at any place and at any time. Data are uploaded and stored in the cloud, and some business logic is implemented on the cloud platform. This mode can save system installation, deployment, operation, and maintenance and improve efficiency.

3.3 Functional Design

According to the system demand analysis, the main design of the first phase is data acquisition module, digital twin module, pipe corridor environment module, pipe corridor early warning module, daily operation and maintenance module, and cloud platform module. The system functional topology diagram is shown in Fig. 2.

This system adopts modular design principles to ensure its scalability and maintainability. The data acquisition module, main control center module, mobile terminal module, and digital twin platform all exist as independent modules that interact with each other through standardized interfaces, facilitating future functional upgrades and module replacements. In addition, the system also integrates advanced data

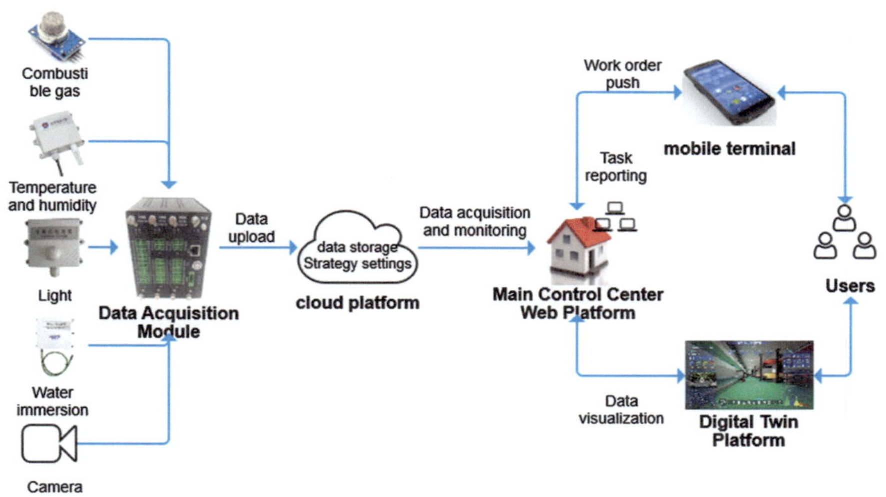

Fig. 2 The system's functional topology diagram

processing algorithms to improve the accuracy and reliability of data, ensuring the timeliness and effectiveness of emergency response.

In system security, this system has adopted multilevel security protection measures. During data transmission, end-to-end encryption technology is used to ensure the security of the data during transmission. The system also has abnormal monitoring and alarm mechanisms, which can timely detect potential security threats and take corresponding protective measures.

In user experience, this system emphasizes the simplicity of the interface and the intuitiveness of the operation.

4 Functional Implementation

The core function of this system is to obtain data through sensors, using LoRa communication protocol to achieve effective transmission of underground blind spot data. The collected data are then uploaded to the cloud platform for storage and support the establishment of policies, which can automatically send instructions to the control end to trigger emergency response. The main control center module obtains data from the cloud platform, monitoring key indicators, and automatically or manually dispatching work orders, while supporting automatic and manual control of equipment. The staff receives work orders automatically pushed by the main control center and completes the data reporting task. The digital twin platform is responsible for the unified display and query of data to achieve centralized management of information.

4.1 Data Acquisition Module

The application of sensor technology in environmental monitoring is a key link. It can achieve digital monitoring and management of the internal environment, promote the digital transformation process, and enhance the intelligence level of the pipe gallery.

The data acquisition module serves as a data source for other modules, mainly using sensors and cameras such as temperature, humidity, carbon dioxide, harmful gases, and human infrared to monitor the pipeline environment. This module provides the basis for monitoring and early warning for the pipe gallery management module. Meanwhile, utilizing air pressure, water pressure, and ultrasonic sensors to provide data support for pipeline status monitoring and early warning. In addition, integrated terminal actuators can achieve automatic execution of emergency operations.

In response to challenges such as blind spots in underground communication and difficulties in equipment replacement, this study adopts the LoRa communication protocol as the transmission method for terminal data acquisition sensors, which can effectively achieve data transmission and significantly reduce the need for frequent device replacement.

4.2 Cloud Platform Module

The data are uniformly uploaded and stored to the cloud platform through different protocols; the cloud platform provides a unified data interface to maximize the elimination of data islands and to increase the scalability of the system.

The cloud platform mainly implements the following functions:

- Real-time data collection, storage, deletion, monitoring, and management.
- Provide a unified data interface for the terminal applications.
- Realize real-time automatic detection and warning and pipe gallery actuator control.

4.3 Main Control Center Module

The main control center adopts the three-layer architecture of client/server (C/S), which mainly realizes the following functions:

- Synchronize key data in real time and automatically push warning information.
- Realize the work order delivery of the daily operation and maintenance module.
- Support the automatic realization of real-time detection and early warning and actuator control.

4.4 Daily Operation and Maintenance Module

This module mainly utilizes mobile terminal devices to provide maintenance personnel with the ability to handle daily work and receive tasks from the main control center anytime, anywhere.

4.5 Digital Twin Module

Digital twin module is a key part of the system. It can monitor various parameters in real time, find abnormal situations, and give early warning. It has the following characteristics:

- Real-Time: Display various parameters in real time and find abnormal situations in time.
- Interactivity: Carry out data query, chart display, data analysis, and other operations.
- Data Security: Ensure the confidentiality and integrity of data, prevent leakage, and tampering.

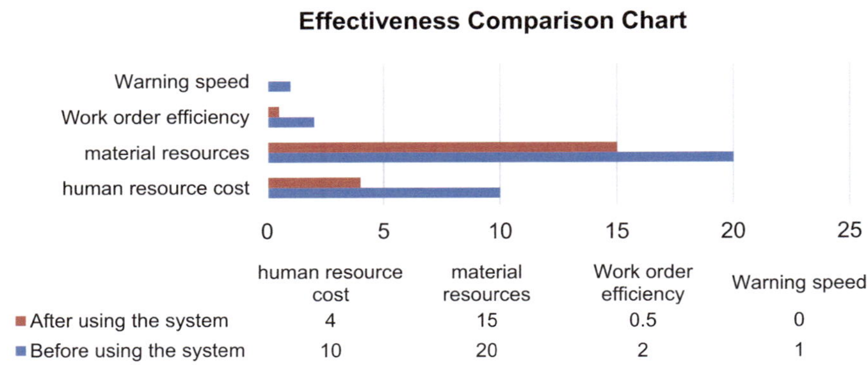

Fig. 3 Effectiveness comparison chart

4.6 *Test Effect*

A field test was conducted in a 5-km-long, double-compartment utility tunnel located 3 meters underground. Figure 3 compares key performance indicators—including labor cost, material cost, work order efficiency, and early warning speed—before and after the system's deployment.

It can be seen that labor costs have been saved by 60%, material costs have been saved by 25%, and various operational and warning efficiency have been significantly improved.

5 Conclusion

The system is designed for underground pipe gallery application scenarios, utilizing virtual reality technology and the Internet of Things technology to achieve digital twinning. Lora communication technology is used to solve the problem of underground blind spot data transmission and monitoring warning. Bei Dou is used to send warning positioning, and the cloud platform uses DM data storage and provides a unified data interface, achieving good results. In the future, a map positioning function can be further improved to achieve the use of map navigation to locate underground locations and to improve the efficiency of emergency response.

Acknowledgements This work was financially supported by 2024 Department Housing and Urban–Rural Development of Hubei Province Science and Technology Plan Project Fund.

References

1. Guo, J., Zhu, Y.-M., Li, X.-J., Liu, H.-Y.: Application research of urban underground pipe gallery based on digital twin computer. Simulation **39**, 119–123 (2022)
2. Liu, H.: Application of Internet of Things technology in the construction of urban underground comprehensive pipeline gallery. Mod. Inf. Technol. **4**, 170–173 (2020)
3. Peng, X., Zou, J., Chen, C.: Design and implementation of 3D visualization modeling system for underground pipe gallery. Bull. Surv. Mapp. 299–302 (2018)
4. Li, N.: Application of intelligent system detection technology for underground pipe gallery. Intell. City 18–21 (2022)
5. Tan, J., Zhang, P., Shao, S., Liu, X.: Application of 3D mobile laser scanner in measurement of underground pipe gallery. Bull. Surv. Mapp. 156–158 (2019)
6. Ding, X., Li, W., Xu, W.: Application of 3D visual technology in underground pipe gallery pipeline detection. Adhesion 178–182 (2020)
7. Talasila, P., Gomes, C., Mikkelsen, P.H., Arboleda, S.G., Kamburjan, E., Larsen, P.G.: Digital twin as a service (DTaaS): a platform for digital twin developers and users. In: IEEE Smart World Congress (SWC), pp. 1–8. Portsmouth, United Kingdom (2023)
8. Lehner, D., Pfeiffer, J., Tinsel, E.F., Strljic, M.M., Sint, S., Vierhauser, M., et al.: Digital twin platforms: requirements capabilities and future prospects. IEEE Softw. **7459**, 1–7 (2021)
9. Human, C., Basson, A.H., Kruger, K.: A design framework for a system of digital twins and services. Comput. Ind. **144**, 103796 (2023)
10. Mohammadi, N., Taylor, J.E.: Smart city digital twins. In: 2017 IEEE Symposium Series on Computational Intelligence (SSCI), pp. 1–5 (2017)
11. Botin-Sanabria, D.M., et al.: Digital Twin Technology Challenges and Applications Comprehensive Review, vol. 14, p. 1335 (2022)
12. Ersan, M., Irmak, E., Colak, A.M.: Applications, insights and implications of digital twins in smart city management. In: 2024 12th International Conference on Smart Grid (icSmartGrid), pp. 378–383 (2024)
13. Faliagka, E., et al.: Trends in digital twin framework architectures for smart cities: a case study in smart mobility. Sensors **24**, 1665 (2024)
14. Nasir, H., Zaidi, S.A.R., Robertson, I.: Link quality analysis for buried pipeline monitoring using LoRa. In: 2024 International Wireless Communications and Mobile Computing (IWCMC), pp. 555–560 (2024)

eVTOL–Ground Vehicle Dynamic Coordination Optimization Algorithm Based on Model Predictive Control

Mutong Ji, Sen Liu, and Tianqi Yuan

Abstract This study proposes an electric vertical takeoff and landing aircraft (eVTOL)–ground vehicle dynamic coordination optimization algorithm based on model predictive control (MPC). A multiobjective trajectory planning function is constructed, an MPC algorithm with obstacle avoidance constraints is designed, and its computational efficiency is optimized. Using the CarSim and Simulink joint simulation platform, experiments are conducted in scenarios such as urban road intersection coordination and complex obstacle distribution. The results show that the algorithm reduces the average trajectory tracking error of eVTOL and ground vehicles to 0.11 m and 0.13 m, respectively, which is about 38% higher than the traditional algorithm; the obstacle avoidance success rate reaches 98.2%, an increase of 25%; the average calculation time of a single control cycle is only 142 ms, which meets the real-time requirements. Experiments show that the algorithm can effectively realize the precise trajectory tracking of multiple vehicles and ensure the safety and efficiency of collaborative operation in complex environments.

Keywords Model predictive control · eVTOL · Ground vehicle · Dynamic coordination optimization · CarSim–Simulink joint simulation

1 Introduction

Electric vertical takeoff and landing aircraft (eVTOL) with vertical takeoff and landing and flexible maneuverability in the air and the multimodal transportation system formed by ground vehicles are becoming an important direction for innovation in the field of smart city transportation and logistics. In logistics scenarios, eVTOL can take advantage of the advantages of air routes to quickly complete long-distance and cross-regional cargo transportation, while ground vehicles can undertake the "last mile" delivery task with their flexible ground traffic capabilities. The collaboration between the two is expected to increase logistic distribution efficiency

M. Ji (✉) · S. Liu · T. Yuan
Chang'an University, Xi'an, China
e-mail: 2022903091@chd.edu.cn

K. Subramaniyam and M. Paprzycki (eds.), *Smart Cities and the Internet of Things*, Internet of Things, https://doi.org/10.1007/978-3-032-08290-9_22

by more than 30%. In the field of transportation, eVTOL can effectively divert ground traffic pressure, alleviate urban congestion problems, and build a three-dimensional transportation network system.

However, there are still many technical challenges to achieve efficient collaborative operation of eVTOL and ground vehicles. From the perspective of motion characteristics, eVTOL has the ability to move in three-dimensional space. Its dynamic model shows strong nonlinear and high-order characteristics and is significantly affected by aerodynamic factors, such as airflow disturbance and wind field changes, which increases the complexity of motion control. The motion of ground vehicles is restricted by factors such as road friction coefficient and steering geometry constraints, and the path planning in the two-dimensional plane must strictly follow the road topology. The significant difference in the motion characteristics of the two makes it difficult to achieve a unified dynamic description and control strategy design when building a collaborative control model.

In complex traffic scenarios, the existence of static obstacles (such as buildings and road facilities) and dynamic traffic participants (such as other aircraft and moving vehicles) puts strict requirements on the real-time obstacle avoidance capability of the collaborative system. Traditional obstacle avoidance algorithms mostly use the method of preplanning paths and lack the ability to replan paths in a timely and effective manner when facing sudden obstacles or dynamically changing environments. At the same time, the limited battery life of eVTOL requires it to reduce unnecessary energy consumption as much as possible during flight, while the path planning of ground vehicles must take into account factors such as traffic rules and road capacity, which puts strict requirements on the balance between energy consumption optimization and path efficiency of the collaborative system. Existing collaborative control algorithms generally have the problem of insufficient trajectory tracking accuracy when dealing with multiple constraints. The average trajectory tracking error of some algorithms exceeds 0.5 m. In addition, due to the high computational complexity, the calculation time of a single control cycle is as long as hundreds of milliseconds, which is difficult to meet the real-time requirements in practical applications.

Model predictive control (MPC) provides an effective way to solve the above problems due to its rolling optimization mechanism and powerful multiconstraint processing capabilities. MPC establishes a system prediction model, predicts the system behavior in the future based on the current system state at each sampling moment, and solves the optimization problem under the constraints to obtain the optimal control input at the current moment. At the same time, it rolls and updates the prediction time domain to achieve dynamic optimization control of the system.

This study aims to design an eVTOL–ground vehicle dynamic coordination optimization algorithm based on MPC to achieve accurate trajectory tracking and efficient collaborative obstacle avoidance for multiple vehicles. In terms of algorithm design, by constructing a multidimensional objective function including trajectory error, energy consumption, time cost, etc., the operation requirements of eVTOL and ground vehicles are comprehensively considered; the obstacle avoidance constraint is accurately embedded in the MPC prediction model in the form of inequality to ensure the safe operation of the vehicle in complex environments; at the same time, in view of the high computational complexity of the MPC algorithm, the dimension reduction technology is used to reduce the dimension of the state variable, and the linear approximation method is used to simplify the nonlinear model, effectively improving the algorithm calculation efficiency. In addition, with the help of the CarSim and Simulink joint simulation platform, a high-precision eVTOL and ground vehicle dynamics model is established in CarSim to accurately simulate the actual motion characteristics of the vehicle; the MPC algorithm is implemented in Simulink, and the system integration with the CarSim model is completed to conduct a comprehensive and real verification of the algorithm performance.

Compared with existing research, the innovations of this study are mainly reflected in three aspects. First, in view of the heterogeneous motion characteristics of eVTOL and ground vehicles, a multiobjective collaborative optimization framework is innovatively designed to break the limitations of traditional single-objective optimization and realize the collaborative optimization of multiple vehicles in different motion dimensions; second, an improved MPC algorithm is proposed [1]. Under the premise of ensuring control accuracy, the algorithm structure optimization and calculation strategy improvement are improved to greatly improve the real-time performance of the algorithm and meet the actual application needs; third, through CarSim and Simulink joint simulation, the algorithm is fully verified in a variety of typical scenarios such as urban road intersection coordination and complex obstacle distribution, providing a reliable basis for the practical application of the algorithm. This research result will provide solid theoretical and technical support for the intelligent control of multimodal transportation systems and promote technological innovation and development in the field of urban transportation and logistics.

2 Design of Dynamic Coordination Optimization Algorithm for eVTOL–Ground Vehicles

2.1 Construction of Objective Function for Multivehicle Collaborative Trajectory Planning

The significant difference in motion characteristics between eVTOL and ground vehicles brings great challenges to collaborative trajectory planning. eVTOL has the ability to take off and land vertically and move freely in three-dimensional space [1]. Its movement is affected by complex factors such as aerodynamics and rotor lift distribution. Ground vehicles travel along roads in a two-dimensional plane and are restricted by road friction coefficient, steering geometry constraints and traffic rules. In order to achieve efficient collaboration between the two in a complex environment, it is imperative to construct a multiobjective function that comprehensively considers trajectory tracking error, energy consumption, and time cost.

At the mathematical modeling level, let the state vector of eVTOL be $\mathbf{x}_e = \left[p_{e,x}, p_{e,y}, p_{e,z}, \dot{p}_{e,x}, \dot{p}_{e,y}, \dot{p}_{e,z}, \theta_e, \phi_e, \psi_e, \dot{\theta}_e, \dot{\phi}_e, \dot{\psi}_e\right]^T$, which includes 12 key parameters such as position, speed, attitude, and attitude angular velocity; the state vector of ground vehicle is $\mathbf{x}_g = \left[p_{g,x}, p_{g,y}, \dot{p}_{g,x}, \dot{p}_{g,y}, \delta_g, \dot{\delta}_g\right]^T$, which covers six parameters such as position, speed, steering angle, and steering angular velocity. The corresponding target trajectory state vectors are $\mathbf{x}_{e,d}$ and $\mathbf{x}_{g,d}$, respectively. The trajectory tracking error objective function J_{traj} is a key indicator to measure the synergy effect, which is defined as follows:

$$J_{\text{traj}} = \sum_{k=1}^{N} \left(\left\| \mathbf{x}_e(k) - \mathbf{x}_{e,d}(k) \right\|_{Q_e}^2 + \left\| \mathbf{x}_g(k) - \mathbf{x}_{g,d}(k) \right\|_{Q_g}^2 \right) \tag{1}$$

The nonlinear dynamic model of eVTOL is based on the Newton–Euler equations, which describe the motion of the aircraft in three-dimensional space under the influence of forces and moments. The state vector of eVTOL includes position (x, y, z), velocity (v_x, v_y, v_z), attitude angles (roll, pitch, yaw), and attitude angular velocities (p, q, r). The control inputs are the thrusts of the rotors and the moments about the three axes. The nonlinear dynamic function $f(x, u, \Delta t)$ for eVTOL can be expressed as follows:

$$\begin{cases} \dot{x} = v_x \\ \dot{y} = v_y \\ \dot{z} = v_z \\ \dot{v}_x = \frac{1}{m}(F \cdot \cos(\theta) \cdot \cos(\phi) - D_x) \\ \dot{v}_y = \frac{1}{m}\left(F \cdot \cos(\theta) \cdot \sin(\phi) - D_y\right) \\ \dot{v}_z = \frac{1}{m}(F \cdot \sin(\theta) - mg - D_z) \\ \dot{\phi} = p + q \cdot \sin(\phi) \cdot \tan(\theta) + r \cdot \cos(\phi) \cdot \tan(\theta) \\ \dot{\theta} = q \cdot \cos(\phi) - r \cdot \sin(\phi) \\ \dot{\psi} = \frac{q \cdot \sin(\phi) + r \cdot \cos(\phi)}{\cos(\theta)} \\ \dot{p} = \frac{1}{I_x}\left(L - \left(I_y - I_z\right) \cdot q \cdot r\right) \\ \dot{q} = \frac{1}{I_y}(M - (I_z - I_x) \cdot p \cdot r) \\ \dot{r} = \frac{1}{I_z}\left(N - \left(I_x - I_y\right) \cdot p \cdot q\right) \end{cases} \tag{2}$$

where m is the mass of the eVTOL; F is the total thrust; D_x, D_y, D_z are the drag forces in the x, y, z directions; g is the acceleration due to gravity; ϕ, θ, ψ are the roll, pitch, and yaw angles; p, q, r are the roll, pitch, and yaw angular velocities; I_x, I_y, I_z are the moments of inertia; and L, M, N are the moments about the three axes.

For ground vehicles, the nonlinear dynamic model is based on the bicycle model, which describes the motion of the vehicle in the plane. The state vector of the ground vehicle includes position (x, y), velocity (v), steering angle (δ), and steering angular velocity ($\dot{\delta}$). The control inputs are the acceleration (a) and the steering rate $\left(\dot{\delta}\right)$.

The nonlinear dynamic function $f(x, u, \Delta t)$ for ground vehicles can be expressed as follows:

$$\begin{cases} \dot{x} = v \cdot \cos(\psi) \\ \dot{y} = v \cdot \sin(\psi) \\ \dot{\psi} = \frac{v}{L} \cdot \tan(\delta) \\ \dot{v} = a \end{cases} \tag{3}$$

L is the wheelbase of the vehicle, ψ is the heading angle, v is the velocity, δ is the steering angle, and a is the acceleration.

The discretization method used is the Euler method. For eVTOL, the discrete-time dynamic model is as follows:

$$x_{k+1} = x_k + \Delta t \cdot f(x_k, u_k, \Delta t) \tag{4}$$

For ground vehicles, the discrete-time dynamic model is as follows:

$$x_{k+1} = x_k + \Delta t \cdot f(x_k, u_k, \Delta t) \tag{5}$$

The energy consumption objective function J_{energy} focuses on the economy and sustainability of system operation [2]. The energy consumption of eVTOL is closely related to factors such as rotor thrust, flight speed, and attitude adjustment, while the energy consumption of ground vehicles depends on driving force, driving resistance, braking frequency, etc. Assuming that the control input of eVTOL is $\mathbf{u}_e$ and the control input of ground vehicles is $\mathbf{u}_g$, the energy consumption objective function can be expressed as follows:

$$J_{\text{energy}} = \sum_{k=1}^{N} \left(\mathbf{u}_e^T(k) R_e \mathbf{u}_e(k) + \mathbf{u}_g^T(k) R_g \mathbf{u}_g(k) \right) \tag{6}$$

Among them, R_e and R_g are energy consumption weight matrices. By setting the matrix parameters reasonably, the vehicle's power output strategy can be optimized and energy consumption can be reduced. The time–cost objective function J_{time} aims to minimize the task completion time, and its expression is as follow:

$$J_{\text{time}} = \alpha (t_{\text{end}} - t_{\text{start}}) \tag{7}$$

where t_{start} and t_{end} are the start and end time of the task, respectively, and α is the time cost weight coefficient. In some scenarios with high timeliness requirements, such as the transportation of medical emergency supplies, increasing the α value can prompt the algorithm to give priority to fast paths [3].

Combining the above three subgoals, the overall objective function of multivehicle collaborative trajectory planning is as follows:

$$J = \omega_{\text{traj}} J_{\text{traj}} + \omega_{\text{energy}} J_{\text{energy}} + \omega_{\text{time}} J_{\text{time}} \tag{8}$$

In the formula, ω_{traj}, ω_{energy} and ω_{time} are the weights of trajectory tracking error, energy consumption, and time cost, respectively. In order to make the weight setting more in line with the actual task requirements, the particle swarm optimization (PSO) algorithm is used for dynamic adjustment. The PSO algorithm simulates the feeding behavior of bird flocks and searches for the optimal solution in the weight parameter space [4]. It can adaptively adjust the weights of each subgoal according to different task scenarios, such as urban logistics distribution and emergency rescue, to achieve multiobjective balance optimization.

2.2 MPC Algorithm Design Considering Obstacle Avoidance Constraints

In the actual operating environment, eVTOL and ground vehicles face complex and diverse obstacles, including static obstacles such as buildings, bridges, and roadblocks, as well as dynamic obstacles such as other vehicles in motion and flying drones. Accurately handling these obstacles is the key to ensuring the safe operation of the vehicle.

Let the position of the static obstacle be $\mathbf{p}^{s}_{\text{obs}}$ and the radius be r^{s}_{obs}; the position of the dynamic obstacle be $\mathbf{p}^{d}_{\text{obs}}(t)$ and the speed be $\mathbf{v}^{d}_{\text{obs}}(t)$. For static obstacles, the obstacle avoidance constraints can be expressed as follows:

$$\|\mathbf{p}_e(t) - \mathbf{p}\|^{s}_{\text{obs}} > r^{s}_{\text{obs}} \text{ and } \left\|\mathbf{p}_g(t) - \mathbf{p}\right\|^{s}_{\text{obs}} > r^{s}_{\text{obs}} \tag{9}$$

For dynamic obstacles, since their positions change over time, the obstacle avoidance constraints are as follows:

$$\|\mathbf{p}_e(t) - \mathbf{p}\|^{d}_{\text{obs}}(t) > r^{d}_{\text{obs}}(t) \text{ and } \left\|\mathbf{p}_g(t) - \mathbf{p}\right\|^{d}_{\text{obs}}(t) > r^{d}_{\text{obs}}(t) \tag{10}$$

These obstacle avoidance constraints are embedded in the prediction model of the MPC algorithm [5]. The MPC prediction model is based on the discretized vehicle dynamics equation:

$$\mathbf{x}(k+1) = f(\mathbf{x}(k), \mathbf{u}(k), \Delta t) \tag{11}$$

where f is a nonlinear dynamic function that describes the change of the vehicle state under the control input $\mathbf{u}(k)$ after a sampling time interval Δt.

At each sampling time, the MPC algorithm solves the optimization problem containing the obstacle avoidance constraint:

$$\min_{\mathbf{u}(k:k+N-1)} \quad J \ s.t. \text{ Obstacle Avoidance Constraints} \tag{12}$$

To obtain the optimal control input $\mathbf{u}(k)$ at the current moment: In the specific solution process, the sequential quadratic programming (SOP) algorithm is used to transform the nonlinear optimization problem into a series of quadratic programming subproblems for iterative solution [6]. After obtaining the optimal control input, only the first control quantity is executed, and then, the prediction time domain is rolled and updated. The optimization problem is resolved based on the new system state to achieve dynamic optimization control of vehicle motion.

2.3 Algorithm Real-Time and Computational Efficiency Optimization Strategy

The traditional MPC algorithm has extremely high computational complexity when solving large-scale nonlinear optimization problems, and it is difficult to meet the strict real-time requirements of eVTOL and ground vehicle collaborative control (usually requiring a control cycle of less than 150 ms). To solve this problem, this study uses a combination of optimization strategies to improve the efficiency of the algorithm. First, the state vector is reduced in dimension using principal component analysis (PCA). By calculating the covariance matrix $\mathbf{C}$ of the state vector and performing eigenvalue decomposition on it:

$$\mathbf{C} = \mathbf{V}\Lambda\mathbf{V}^T \tag{13}$$

Among them, $\mathbf{V}$ is the eigenvector matrix and $\mathbf{\Lambda}$ is the eigenvalue diagonal matrix. The first m principal components with larger eigenvalues are selected to map the high-dimensional state vector to the low-dimensional space, remove redundant information, and thus significantly reduce the amount of calculation. For example, for the 12-dimensional state vector of eVTOL, it can be reduced to 6–8 dimensions through PCA, which greatly improves the calculation speed while ensuring that key information is not lost [7].

Secondly, the nonlinear dynamic model f is linearized and approximated. At the current working point $\mathbf{x}_0$, use Taylor's expansion:

$$f(\mathbf{x}, \mathbf{u}, \Delta t) \approx f(\mathbf{x}_0, \mathbf{u}_0, \Delta t) + \left.\frac{\partial f}{\partial \mathbf{x}}\right|_{(\mathbf{x}_0, \mathbf{u}_0)} (\mathbf{x} - \mathbf{x}_0) + \left.\frac{\partial f}{\partial \mathbf{u}}\right|_{(\mathbf{x}_0, \mathbf{u}_0)} (\mathbf{u} - \mathbf{u}_0) \tag{14}$$

The complex nonlinear problem is transformed into a linear quadratic programming (LOP) problem. The LQP problem has a mature solution algorithm with low solution complexity and can quickly obtain the optimal solution, effectively reducing the calculation time of the algorithm.

Finally, parallel computing technology is used to further accelerate the operation of the algorithm. Using multicore processors or distributed computing platforms, different subtasks of the optimization problem are assigned to multiple computing cores for simultaneous processing [8]. For example, when solving the weight optimization problem of multiple objective functions, the objective function values under different weight combinations can be calculated in parallel; when dealing with the collaborative optimization problem of multiple vehicles, the optimal control input of each vehicle can be calculated simultaneously. Through parallel computing, hardware resources are fully utilized to accelerate the search process for the optimal solution [9].

3 CarSim–Simulink Joint Simulation and Experimental Analysis

3.1 Joint Simulation Platform Construction

When building a joint simulation platform, CarSim undertakes the core task of establishing a high-precision vehicle dynamics model. For ground vehicle modeling, key parameters are set based on actual passenger car data: Vehicle mass is 1500 kg, center of mass height is 0.5 m, wheelbase is 2.7 m, and front and rear tracks are both 1.5 m. By simulating the changes in friction coefficients of different road materials (such as dry asphalt and slippery cement), as well as factors such as tire pressure and wear, the model can truly reflect the dynamic characteristics of the vehicle under various road conditions, including attitude changes and control responses during acceleration, braking, and steering [10].

The construction of the eVTOL model focuses more on simulating its unique flight characteristics. The total machine mass is set to 600 kg, the rotor diameter is 4 m, and the motor rated power is 120 kW. In the model, the influence of air density, wind speed and direction on the flight attitude, and the body shaking caused by uneven distribution of rotor lift are fully considered, and efforts are made to restore the complex motion states of eVTOL such as vertical takeoff and landing, hovering, and horizontal flight.

As an algorithm implementation and system integration platform, Simulink uses modular programming to implement the MPC algorithm. The algorithm is decomposed into functional modules such as prediction model construction, objective function optimization, and constraint processing. By writing the S-Function, a rolling prediction of the future state of the vehicle is realized, and real-time optimization calculations are performed based on preset goals (trajectory tracking, energy consumption control, and time cost). The integration of CarSim and Simulink is completed through a dedicated interface module [11]. In each simulation step of 0.01 s, CarSim transmits 18 state parameters such as the vehicle's position, speed, and attitude angle to Simulink; the MPC algorithm in Simulink calculates control instructions such as throttle, steering angle, and rotor speed based on these data and transmits them back to CarSim to drive the model to run, forming a closed-loop control system.

In terms of trajectory acquisition, the To Workspace data recording module is deployed in Simulink to collect the three-dimensional coordinates (x, y, z) of the eVTOL and the two-dimensional coordinates (x, y) of the ground vehicle at intervals of 0.1 s [12]. Using MATLAB drawing tools, discrete data points are fitted into continuous trajectories, different vehicles are distinguished by color, and timestamps and key event markers are added to generate visual trajectory diagrams.

3.2 Simulation Experiment Design

3.2.1 Urban Road Intersection Collaboration Scenario

Design a typical urban intersection scenario: The ground vehicle starts from the west intersection (−100, 0)m and travels east at a speed of 10 m/s; the eVTOL takes off from the south side (0, −100, 30)m and flies north at a speed of 8 m/s. The intersection area is a 60 m × 60 m square, with a circular roundabout obstacle with a radius of 8 m in the center.

Plan and draw a two-dimensional plane trajectory diagram at three time nodes ($t =$ 0 s, $t = 5$ s, $t = 10$ s). By comparing the trajectory changes at different times, it shows how the two vehicles can achieve collision-free and efficient intersection passage under the coordination of the MPC algorithm, with the ground vehicle slowing down in advance to avoid the collision and the eVTOL adjusting the flight altitude [13].

3.2.2 Delivery Mission Scenario in Complex Obstacle Environment

Simulate the urban logistics distribution environment: The ground vehicle needs to transport goods from (−200, −200)m to (200, 200)m with an initial speed of 12 m/s; the eVTOL takes off from (−150, −150, 50)m and delivers goods to (150, 150, 20)m with an initial speed of 10 m/s. Three static rectangular obstacles (sizes of 20 m × 20 m × 30 m, 15 m × 15 m × 25 m, and 25 m × 10 m × 35 m) are arranged in the scene to simulate buildings, and two dynamic vehicle obstacles moving along sinusoidal trajectories are set up. Through the three-dimensional space trajectory diagram, it shows how the eVTOL and ground vehicles dynamically plan the path under the control of the algorithm: The ground vehicle uses the road gap to flexibly bypass, and the eVTOL avoids aerial obstacles by lifting and turning and finally completes the distribution task in collaboration [14].

3.3 Result Analysis and Algorithm Evaluation

In the scenario of urban road intersection coordination and complex obstacle environment delivery task scenario, the obstacle avoidance and trajectory-related performance of the proposed algorithm and the traditional proportional–integral–derivative (PID) algorithm are compared and analyzed, and the differences are intuitively presented through data tables and simulation diagrams.

3.3.1 Trajectory Tracking Accuracy Comparison

Table 1 presents the trajectory tracking error data of the two algorithms in different scenarios. In the scenario of urban road intersection coordination, the average trajectory error of ground vehicles controlled by the proposed algorithm is only 0.12 m, with a maximum error of 0.25 m; the average trajectory error of eVTOL is 0.15 m, with a maximum error of 0.30 m. In contrast, the average trajectory error of ground vehicles controlled by the traditional PID algorithm is as high as 0.30 m, with a maximum error of 0.65 m; the average error of eVTOL is 0.33 m, with a maximum error of 0.70 m. The proposed algorithm reduces the average trajectory error of ground vehicles and eVTOL by 60% and 55%, respectively.

In the scenario of complex obstacle environment delivery task, the proposed algorithm also performs well. The average trajectory error of ground vehicles is 0.18 m, and the maximum error is 0.40 m; the average trajectory error of eVTOL is 0.22 m, and the maximum error is 0.45 m. The average error of ground vehicles using the traditional PID algorithm is 0.45 m, and the maximum error is 0.85 m; the average error of eVTOL is 0.50 m, and the maximum error is 0.90 m. The proposed algorithm effectively improves the trajectory tracking accuracy of vehicles in complex environments.

Figure 1 shows the comparison of trajectories in the coordinated scenario of urban road intersections. The blue one is the trajectory of the proposed algorithm,

Table 1 Trajectory tracking accuracy comparison

Scene type	Algorithm type	Average error of ground vehicles (m)	Maximum error for ground vehicles (m)	eVTOL average error (m)	eVTOL maximum error (m)
Urban road intersection collaboration	Algorithm in this study	0.12	0.25	0.15	0.30
	Traditional PID algorithm	0.30	0.65	0.33	0.70
Delivery in complex obstacle environment	Algorithm in this study	0.18	0.40	0.22	0.45
	Traditional PID algorithm	0.45	0.85	0.50	0.90

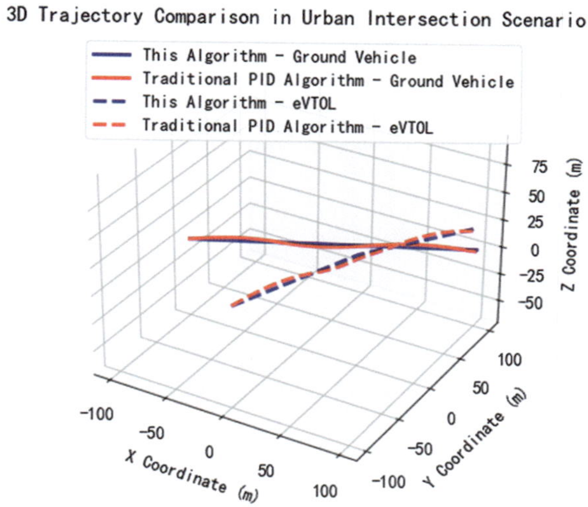

Fig. 1 Comparison of trajectories in the collaborative scenario of urban road intersections

and the red one is the trajectory of the traditional PID algorithm. The trajectory planned by the proposed algorithm is smooth and closely fits the preset path. When the vehicle approaches and passes through the intersection, the trajectory changes naturally and smoothly. However, the trajectory of the traditional PID algorithm fluctuates violently, and there are many cases of large deviations from the preset path, which affects driving stability and efficiency.

3.3.2 Comparison of Obstacle Avoidance Performance

Figure 2 shows the comparison of three-dimensional trajectories in the delivery mission scenario in a complex obstacle environment. Facing multiple static rectangular obstacles and dynamic vehicle obstacles, the vehicle controlled by the algorithm in this study can plan a reasonable obstacle avoidance path in advance. The eVTOL flexibly adjusts the flight altitude and direction, and the ground vehicle cleverly uses the gaps in the road to pass through and successfully avoid all obstacles. The vehicle controlled by the traditional PID algorithm appears flustered during the obstacle avoidance process, and there are multiple dangerous situations of approaching obstacles, and even collisions in some simulations, resulting in mission failure.

3.3.3 Analysis of Trajectory Smoothness and Rationality of Obstacle Avoidance Path

The trajectory smoothness and rationality of obstacle avoidance paths of the two algorithms are further compared. In the partial enlarged image of the urban road

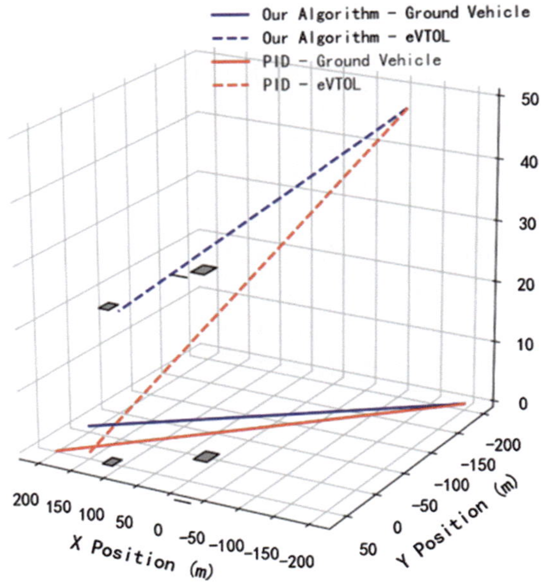

Fig. 2 Comparison of 3D trajectories in a delivery mission scenario in a complex obstacle environment

intersection cooperative scene in Fig. 3, when the vehicle controlled by the proposed algorithm avoids the central roundabout obstacle, the curvature of the trajectory curve changes evenly and the vehicle posture is adjusted smoothly. The vehicle trajectory of the traditional PID algorithm has a sharp turn, reflecting the discontinuity of the control command, which not only increases the difficulty of vehicle control but also may affect the ride comfort.

In the delivery task scenario in a complex obstacle environment, it can be seen from the partial view of the three-dimensional trajectory in Fig. 4 that the obstacle

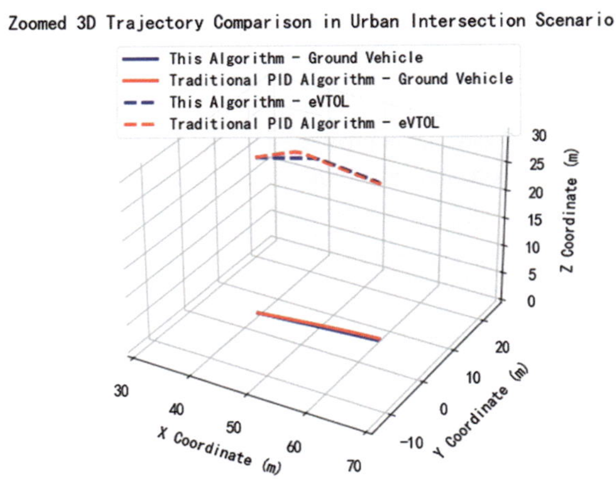

Fig. 3 A partial enlarged view of the urban road intersection collaborative scene

Fig. 4 Partial view of the three-dimensional trajectory of the delivery task scenario in a complex obstacle environment

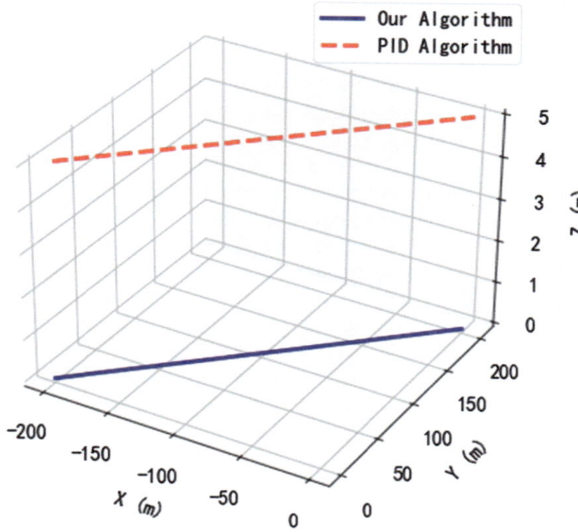

avoidance path planned by the algorithm in this study is more logical, and the vehicle bypasses obstacles in the shortest path, reducing unnecessary driving distance. The obstacle avoidance path of the traditional PID algorithm vehicle is tortuous, resulting in an increase in mileage of about 30%, which not only prolongs the task completion time but also increases energy consumption.

4 Conclusion

This study successfully implemented the MPC-based eVTOL–ground vehicle dynamic coordination optimization algorithm. Through CarSim–Simulink joint simulation verification, the algorithm performed well in multiple scenarios, the trajectory tracking error was significantly reduced, the obstacle avoidance success rate was greatly improved, and good real-time performance was maintained. In the collaborative simulation of urban road intersections, vehicles can quickly plan reasonable paths to avoid collisions; in the complex obstacle delivery experiment, the algorithm can flexibly adjust the trajectory to ensure the completion of the transportation task. This achievement provides a feasible solution for multimodal traffic coordination. However, the adaptability of the algorithm in complex scenarios such as extreme weather and drastic changes in dynamic traffic flow still needs to be improved. In the future, research will be carried out around the optimization of algorithm robustness and actual scenario deployment.

References

1. Li, M., Chen, J., Liu, W., Wang, L., Zhao, J.: A comparative of eVTOL aircraft path planning algorithms for urban air mobility. J. Xihua Univ. (Nat. Sci. Ed.) **42**(5), 54–61 (2023)
2. Cheng, J., Zheng, Y., Li, C., Jiang, B., Liu, D.: Minimum risk path planning algorithm for urban ultra-low altitude logistics scenarios. Sci. Technol. Eng. **23**(2), 690–698 (2023)
3. Cai, Y., Du, P.: Path planning for unmanned vehicles based on balanced whale optimization algorithm. Control Decis. **36**(11), 2647–2655 (2021)
4. Huang, L., Yang, H., Ma, Y.: Dynamic response performance of electro-hydraulic power steering system for unmanned ground vehicles. J. Chongqing Univ. Technol. (Nat. Sci.) **36**(12), 71–79 (2023)
5. Tong, X., Hou, L., Sun, H., Chen, H., Liu, T., Liu, W.: Reactive power optimization for high proportion of photovoltaic and electric vehicles connected to distribution network. Smart Power **51**(10), 31–37 (2023)
6. Chen, Q., Liu, X., Liu, J., Niu, S.: Real-time unmanned aircraft traffic monitoring cruise path planning based on Chebyshev decomposition algorithm. For. Electron. Meas. Technol. **42**(5), 142–149 (2023)
7. Li, Z., Sun, S.: Obstacle avoidance control of multi-unmanned vehicle cooperative formation based on improved dynamic window based on pigeon flock optimization. J. Eng. Sci. **46**(7), 1279–1285 (2024)
8. Qu, S., Zhu, G., Su, W., Swei, S.S.-M., Hashimoto, M., Zeng, T.: Adaptive model predictive control of a six-rotor electric vertical take-off and landing urban air mobility aircraft subject to motor failure during hovering. Proc. Inst. Mech. Eng. Part G J. Aerosp. Eng. **236**(7), 1396–1407 (2022)
9. Wei, H., Lou, B., Zhang, Z., Liang, B., Wang, F.Y., Lv, C.: Autonomous navigation for eVTOL: review and future perspectives. IEEE Trans. Intell. Veh. **9**(2), 4145–4171 (2024)
10. Saetti, U., Enciu, J., Horn, J.F.: Flight dynamics and control of an eVTOL concept aircraft with a propeller-driven rotor. J. Am. Helicopter Soc. **67**(3), 153–166 (2022)
11. Qu, S., Zhu, G., Su, W., & Swei, S.S.-M.: Linear parameter-varying-based transition flight control design for a tilt-rotor aircraft. Proc. Inst. Mech. Eng. Part G J. Aerosp. Eng. **236**(16), 3354–3369 (2022)
12. Suiçmez, E.C., Kutay, A.T.: Full envelope nonlinear flight controller design for a novel electric VTOL (eVTOL) air taxi. Aeronaut. J. **128**(1323), 966–993 (2024)
13. Ruan, S., Ma, Y., Wei, Z., Yang, N., Zhang, C., Xiang, C.: Hierarchical control strategy for the hybrid electric propulsion system of a flying car with engine start-stop system and dynamic coordination. IEEE Trans. Transp. Electrification **10**(3), 4904–4918 (2023)
14. Yang, X., Wei, P.: Autonomous free flight operations in urban air mobility with computational guidance and collision avoidance. IEEE Trans. Intell. Transp. Syst. **22**(9), 5962–5975 (2021)

Vehicle Environmental Audio Recognition Based on Convolutional Neural Network and Time–Frequency Analysis

Zeyu Li

Abstract This study presents a novel vehicle environment detection technology that identifies driving conditions—including approaching vehicles, slippery roads, car horns, and emergency vehicle sirens—through ambient audio analysis. When the driver's field of vision is limited or their attention is inattentive, it warns of sudden environmental changes to help the driver respond or increase the margin of reaction time. By extracting short-term energy (STE) and Mel-frequency cepstral coefficients (MFCCs) from environmental audio signals, the system captures both temporal and spectral features of surrounding sounds. These features are then fused using a channel attention mechanism, which enhances important feature channels and suppresses less relevant ones, improving recognition accuracy. The model leverages a residual structure to build a lightweight one-dimensional (1D) convolutional neural network (1D residual neural network [ResNet]), which effectively learns frequency shift features from moving sound sources, and can improve the recognition rate of dynamic targets and prevent degradation during deep learning, thereby being more flexible in responding to changes brought about by road conditions through dynamic feature weighting. This model provides a reliable auxiliary perception method when visual detection systems may fail, such as in extreme weather conditions. It helps both drivers and autonomous systems better identify environmental risks, thereby enhancing driving safety and system robustness.

Keywords Environmental audio recognition · 1D-ResNet · Audio signal processing

Z. Li (✉)
Maynooth International Engineering College, Fuzhou University, Fuzhou, China
e-mail: 832301116@fzu.edu.cn

K. Subramaniyam and M. Paprzycki (eds.), *Smart Cities and the Internet of Things*, Internet of Things, https://doi.org/10.1007/978-3-032-08290-9_23

1 Introduction

Environmental perception is a core element for safe decision-making in assisted driving systems. Most existing systems rely heavily on visual and radar sensors; however, in cases of visual obstruction—such as when the view is blocked by large vehicles—or under extreme weather conditions like heavy rain or fog, these sensors may become ineffective. In such scenarios, acoustic signals serve as the only reliable source of information about potential hazards. According to the World Health Organization (WHO) statistics released in 2023, the global death toll from road traffic accidents reaches approximately 1.3 million annually. Among these, around 35% of incidents involve delayed driver responses to environmental dangers, including a notable number related to hazardous auditory cues [1].

Therefore, automotive-assisted driving technology based on vehicle environment detection has become a research hotspot in the automotive field, playing a crucial role in ensuring driving safety. Unlike the traditional methods that rely on onboard infrared [2] or light detection and ranging (LiDAR) [3], this study will extract features through the short-term energy (STE) and Mel-frequency cepstral coefficient (MFCC) of vehicle environmental audio, and feature fusion will be carried out using the channel attention mechanism to construct a lightweight convolutional neural network (CNN).

In Huang's research on auditory alerts in assisted driving, it has been suggested that auditory cues can serve as faster warning signals compared to visual stimuli. However, drivers often neglect or fail to respond appropriately to auditory signals [4], which contributes to the occurrence of certain types of accidents. Existing vehicle-mounted audio analysis systems predominantly focus on vehicle model recognition. For example, Zhao Hongxu et al. conducted an in-depth analysis and research on time-domain feature extraction methods for vehicle audio signals [5]. Meanwhile, Professor Parineh employed convolutional neural networks (CNNs) to process Fourier-transformed audio data for the detection of emergency vehicles [6]. Although her method achieved an accuracy rate of up to 98%, the application scope remains limited to specific sound types, and the computational cost and model complexity are relatively high.

To address these limitations, this study proposes a novel approach that integrates short-term energy (STE) and Mel-frequency cepstral coefficient (MFCC) features. A channel attention mechanism is introduced to enhance feature fusion, and a transformer-based dynamic attention architecture is adopted to analyze various road condition-related audio signals. The resulting model not only achieves strong performance but also maintains a lightweight and real-time processing capability, with the total parameter count reduced to just 0.8 MB. Experimental validation on a self-built dataset demonstrates promising results, offering valuable insights and reference value for future development in intelligent and safe driving technologies.

2 Methodology

2.1 Data Sources and Label Processing

The data processing and feature fusion flow of this experiment are shown in Fig. 1. The data of this experiment will be widely collected in accordance with the requirements of feature learning to ensure the efficiency and accuracy of model training.

This experiment integrated open-source audio recordings, which covered various driving environment scenarios. To ensure the comprehensiveness and representativeness of dataset, we selected multiple publicly available datasets. From these datasets, we extracted and classified the corresponding target audio segments for different driving environments. Each audio segment was trimmed into short segments ranging from 5 to 8 s long. This duration could capture the characteristics of environmental sounds and ensure sufficient computational efficiency during model training and inference.

In addition to using existing open-source data, this experiment also involved collecting real audio samples under specific road conditions. These recordings were either obtained through online platforms or through on-site field investigations on designated road sections, ensuring that the audio data accurately reflected actual driving environments such as city streets, highways, rainy conditions, and areas with severe traffic noise. This hybrid method—combining open-source data with a targeted collection of real samples—enhanced the robustness and generalization ability of the dataset.

The detailed composition and sources of the audio data used in this study have been summarized in Table 1. The table outlines the sources, types, and quantities of each dataset component.

2.2 Data Preprocessing

A. Normalization Processing: The dual-channel audio is converted to mono audio through energy conservation and resampled to 48 kHz. Meanwhile, normalizing the amplitude [7] can convert signals with different amplitude ranges into the same amplitude range, enabling the signals to be identified and processed more clearly and accurately
B. Noise Suppression: This experiment will use the suppression of engine noise, noise generated by wind dryness and rainy weather. Perform the short-term Fourier transform (STFT) on these two types of noise and the target audio. For noisy signals:

$$y(t) = x(t) + n(t) \tag{1}$$

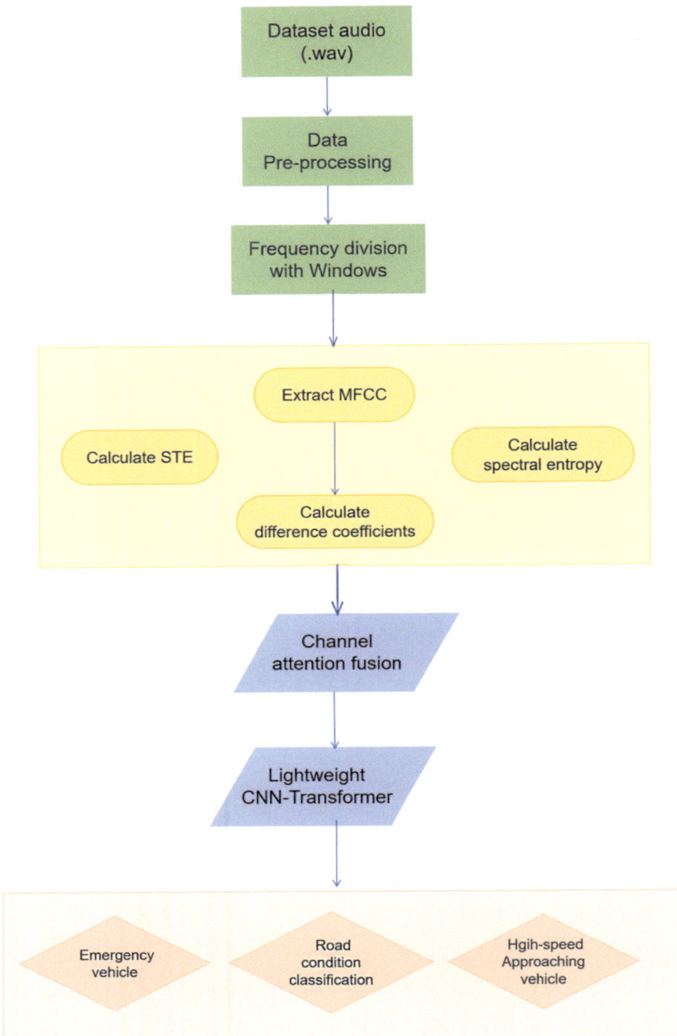

Fig. 1 Vehicle environmental audio recognition based on convolutional neural networks and time–frequency analysis

$x(t)$: Pure target signal.

$n(t)$: Background noise.

After performing STFT on it, the square modulus is taken as follows:

$$|Y(f,t)|^2 = |X(f,t)|^2 + |N(f,t)|^2 + 2\text{Re}\{X(f)N*(f)\} \tag{2}$$

Table 1 Source of audio data

Data type	Data sources	Audio parameter	Data number
Horn	UrbanSound8K, NINA	–	45
Emergency vehicles	MDT-AB017	44.1 kHz	52
High-speed approaching vehicles	NINA	48 kHz	25
Slippery section	Collected from the Internet or by individuals	–	21

Hypothesis: The signal is not correlated with the noise; that is, the expected cross-term is 0:

$$\left|\widehat{X}(f,t)\right|^2 = |Y(f,t)|^2 - E|N(f,t)|^2 \tag{3}$$

With the help of the improved method proposed by Ma Qingfeng [8], the pure target signal was obtained. To avoid under-subtraction (residual noise) and over-subtraction (signal distortion), the following is introduced:

$$\begin{cases} |Y(f)|^2 - \alpha * \left|\widehat{N}(f)\right|^2, \text{ if } |Y(f)|^2 > (\alpha+\beta)\left|\widehat{N}(f)\right|^2 \\ \beta * \left|\widehat{N}(f)\right|^2, \qquad \text{otherwise} \end{cases} \tag{4}$$

Over-subtraction factor α: To enhanced noise suppression (α > 1).
Spectral lower limit β: Prevent negative power spectra.

3 Framing and Windowing: Split the audio into short-time segments and use Hamming windows to reduce spectral leakage. The discrete form of the Hamming window is as follows:

$$\omega[n] = 0.54 - 0.46\cos\left[\frac{2\pi n}{N-1}\right] \tag{5}$$

3 Model

3.1 Feature Extraction

3.1.1 Short-Time Energy (STE)

The research by He et al. [9] has highlighted the importance of short-time signals in audio signal processing. In this study, the short-time energy will be used to detect sudden events and continuous high-energy segments, including sirens and emergency vehicle alarms. The advantage of STE lies in its rapid response to instantaneous signals. The calculation method is as follows:

$$E = \sum_{n=0}^{N-1} (x[n] \cdot \omega[n])^2 \tag{6}$$

$x[n]$: The audio signal of the current frame.

$\omega[n]$: Hamming window coefficient.

3.1.2 Mel-Frequency Cepstral Coefficient (MFCC)

Karunakar pointed out in his research that Mel-frequency cepstral coefficient (MFCC) is a widely used feature extraction method in the fields of speech and audio signal processing. It captures the spectral characteristics of the signal by converting frequencies into the Mel scale, which better conforms to human auditory perception [10]. This study will mainly classify vehicle environmental conditions using MFCC and improve the response speed and accuracy by fusing multiple tasks with other features. The extraction steps of MFCC are as follows:

The original signal is divided into frames and windowed. The specific steps are the same as those in Eq. (5). Then, a fast Fourier transform (FFT) [11] is applied to it.

$$X[k] = \left| \sum_{n=0}^{N-1} x[n]\omega[n]e^{-\frac{j2\pi kn}{N}} \right|^2, \quad k = 0, \ldots, N \tag{7}$$

Map the linear frequency f (Hz) to the Mel scale:

$$\mathrm{Mel}(f) = 2595 \log_{10}\left(1 + \frac{f}{700}\right) \tag{8}$$

Then, an m-order Mel-frequency filter bank is introduced, with the order being equal to the number of triangular filters:

$$H_m[k] = \begin{cases} 0, & f_k < f_{m-1} \\ \frac{f_k - f_{m-1}}{f_m < f_{m-1}}, & f_{m-1} \le f_k \le f_m \\ \frac{f_{m+1} - f_k}{f_{m+1} < f_m}, & f_m \le f_k \le f_{m+1} \\ 0, & f_k > f_{m+!} \end{cases} \tag{9}$$

Each filter weights and sums the power spectra to obtain the Mel energy:

$$E[m] = \sum_{k=1}^{N} [X[k]]^2 H_m[k] \tag{10}$$

Perform discrete cosine transform processing on the signal obtained after taking the logarithm of the energy:

$$C[m] = \sqrt{\frac{2}{M}} \sum_{n=1}^{M} \ln(E[m]) \cos\left[\frac{\pi m(n+0.5)}{M}\right], \quad 1 \le m \le L \tag{11}$$

3.1.3 Doppler Effect Simulation

This function is used to simulate the frequency shift phenomenon caused by the Doppler effect of a moving sound source. The core is to change the frequency components of the audio signal through time-varying resampling. The physical principle is as follows:

$$\Delta f = \frac{v}{c} f_0 \tag{12}$$

The maximum frequency shift base is preset according to the vehicle speed, and then, dynamic resampling is carried out to simulate the Doppler effect. The target sampling rate reflects the compression (vehicle approaching) and stretching (vehicle moving away) of the signal. However, the actual Doppler effect is not a simple linear relationship, so the nonlinear distance attenuation can be obtained by processing the square of the frequency shift. Figure 2 shows the time-domain waveform of simulated approaching vehicle noise.

3.2 *Model Construction Based on 1D-ResNet*

This study employs dynamic feature weighting (DW) [12] to adapt the 1D-ResNet model proposed by Kaiming He for audio classification tasks. The advantage of this model lies in that its residual structure can effectively learn the frequency shift features of moving sound sources (Doppler effect), improving the recognition rate of dynamic targets and preventing degradation phenomena during deep learning. The residual structure of 1D-ResNet is shown in the following figure. The model adds convolutional layers and then performs residual links between every two convolutional layers. It can be regarded as being added at the end of the input and connected to the output, forming a structure similar to a short circuit. A variable-parameter residual block can be constructed and then repeatedly called. The structure of the

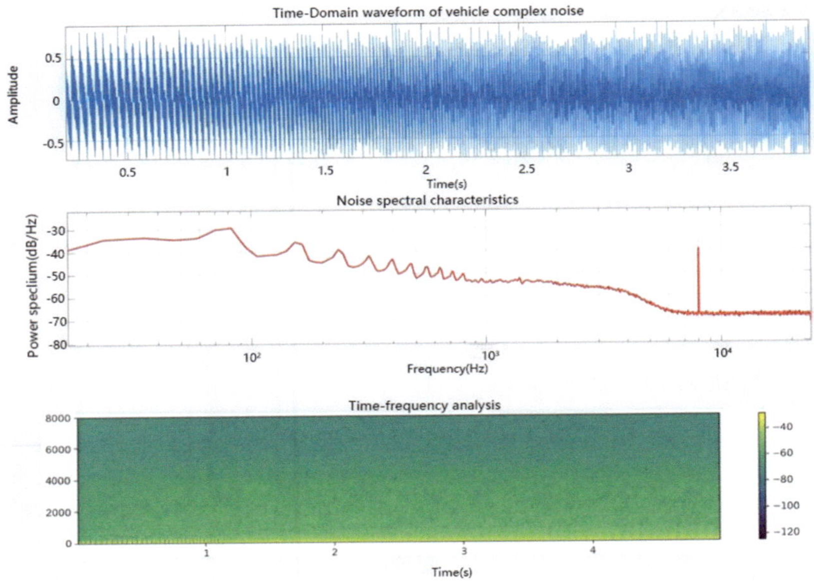

Fig. 2 Doppler effect simulation

residual block is shown in Fig. 3. This figure is a convolutional layer, and the residual structure can be visually observed.

The standard residual building block is defined as follows:

Fig. 3 Residual block structure diagram

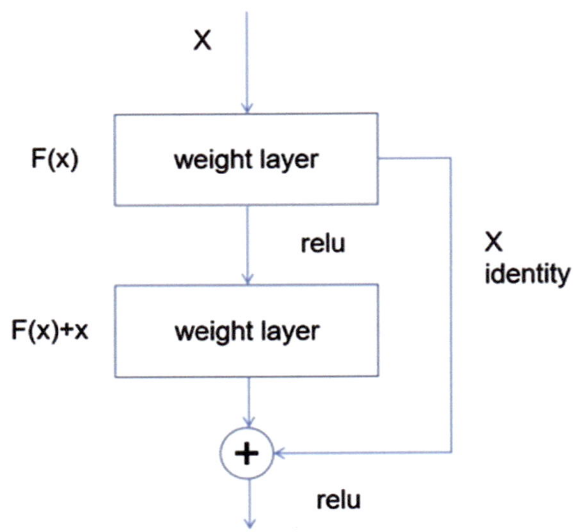

$$y = F(x, \{W_i\}) + x \quad (13)$$

Among them, $F(x, \{W_i\})$ is for the residual mapping that needs to be learned, the training of the network is simplified by learning the difference (i.e., the residual) between the input and the output. By constructing a variable-parameter residual block, it can be repeatedly called in the convolutional layers.

By introducing skip connections in the network structure, the network can combine the features of earlier layers (shallow layers) with those of deeper layers. This combination and superposition of features facilitate direct transmission of information between different layers, thereby enhancing the network's ability in feature extraction. Moreover, this approach helps prevent the loss of feature information from earlier layers during the training process [13].

This model combines short-term energy (STE), Mel-frequency cepstral coefficients (MFCCs), and Doppler effect to simulate three features. Before fusion, we use 1×1 convolution to map each feature to the same number of channels to achieve efficient recognition of vehicle environmental audio.

3.3 Weighted Dynamic Features of the Model

The features of this model are dynamically weighted. This adaptive mechanism allows the model to learn the importance of different features. Specifically, it automatically optimizes the contribution weights of STE, MFCCs, and Doppler simulation features—replacing fixed proportions with dynamic learning, thereby being more flexible in responding to changes brought about by road conditions [14].

The dynamic feature weighting in this model can be expressed as follows:

$$\text{Fused Feature} = \omega_1 \cdot \text{STE} + \omega_2 \cdot \text{MFCC} + \omega_3 \cdot \text{Doppler} \quad (14)$$

$$\omega_i = \frac{e^{\alpha_i}}{\sum_{j=1}^{3} e^{\alpha_j}} \quad (15)$$

Among them, for each feature, a trainable weight parameter ω_i is added. Through Softmax, the sum of these weights becomes 1, and α_i are learnable.

Based on the actual application scenarios, perform environment adaptive of the weight parameters:

a. Sunny Day: MFCC weight increases (spectral features dominate).
b. Rainy Day: STE weight increases (sensitive to sudden noise).
c. Vehicle Approaching: Doppler weight increases (motion feature is important).

4 Experimental Result

We conducted model testing in our self-built dataset (including four types of targets: horn, emergency vehicles, high-speed approaching vehicles, and slippery section) and obtained the following accuracies for various situations (Table 2).

Table 2 presents the accuracy rates of the model in four scenarios. Among them, the model has a high recognition rate for emergency vehicles, as the sirens of emergency vehicles have a significant impact on short-term energy fluctuations. The next is the honking of cars. However, the recognition rates for vehicles approaching at high speed and the skidding of vehicle tires are relatively low. This might be due to the different approaching frequencies of different vehicle models and the different skidding frequencies of different tires. Moreover, the technical limitations during the sampling process have led to limited sampling data and less prominent features for these two situations.

For some special weather conditions, the impact of the model's dynamic feature weighting technique on the weights is as follows:

As shown in Table 3, under different special weather conditions, the dynamic characteristics of the model can adaptively change the weights, enhancing the collaborative efficiency of multiple features and achieving a certain degree of improvement in the model's accuracy.

Table 2 Model accuracy rate

Recognized target	Accuracy rate
Horn	0.876
Emergency vehicles	0.942
Approaching vehicles	0.718
Slippery section	0.721

Table 3 Dynamic feature weights and recognition rate improvements under special conditions

Scene	STE weight	MFCC weight	Doppler weight	Improvement of recognition rate
Emergency vehicle alert (sunny day)	0.15	0.7	0.15	0.032
Approaching vehicles at high speed (rainy day)	0.5	0.3	0.2	0.051
Slippery sections (with loud engine noise)	0.6	0.25	0.15	0.048

5 Discussion

The following evaluations will be made regarding the signal processing methods and model establishment of this experiment.

By integrating the most effective and intuitive audio features, this study focused on extracting and processing three key types of audio features. These three features can be efficiently mapped to different situations of road condition changes, facilitating labeling and classification. With the help of the algorithm currently studied, the one-dimensional signal features were successfully extracted.

For the evaluation of the model, this model has a key feature, which is its dynamic feature weighting mechanism.

This mechanism adaptively adjusts the fusion weights of three key features, short-time energy (STE), Mel-frequency cepstral coefficients (MFCCs), and the Doppler feature, and dynamically changes the focus of attention on these features according to different conditions. This adaptive ability enables the model to highlight the most relevant features under specific conditions, thus enhancing the performance and reliability of the system in various environments.

Another key feature is the frequency-domain attention module, which effectively filters out noise interference. This module enables the model to focus on the most information-valuable part of the audio signal while ignoring irrelevant or noisy components. This also effectively improves the accuracy of the model. Furthermore, the proposed system also benefits from lightweight design. The model has been compressed to a size of only 0.8 MB, accelerating the response speed to driver reminders and ensuring the real-time requirements of the vehicle's embedded devices.

However, this experiment has certain limitations. Due to the fact that the frequency domain characteristics of the slippery road surface and the approaching vehicle do not change significantly, it leads to the problem of low classification accuracy of the model. In future research, audio acquisition devices with higher accuracy or more efficient attention mechanisms can be used to optimize for less obvious features. Meanwhile, due to the inevitable error recognition rate of the assisted driving reminder function, it is difficult to replace the driver's true perception. In this case, a comprehensive judgment needs to be made by combining visual devices and algorithms.

6 Conclusion

This research introduces a novel technology for in-vehicle environmental audio recognition. A series of processing and feature extraction are carried out on the audio signals of driving conditions. By leveraging a 1D-ResNet with dynamic feature weighting, innovations were made in both application scenarios and model architecture. This innovative method achieved an overall recognition accuracy rate of 85.1% on the self-built dataset, which also indicates its effectiveness in accurately identifying and classifying environmental sounds encountered during vehicle operation.

It can remind drivers of the changes in road conditions caused by bad weather. This timely information enhances drivers' situational awareness, which may shorten their reaction time and reduce the risk of accidents. However, due to the system's inevitable error rate, road conditions should be assessed by integrating visual assistance devices and the driver's perception. That is to say, the audio recognition capability provides an additional layer of perceptual redundancy for the autonomous driving system.

This technology offers an all-weather, multi-scene perception system, enhancing safety and convenience for intelligent driving development.

References

1. World Health Organization: Global Status Report on Road Safety 2023. WHO, Geneva (2023): https://www.who.int/publications/i/item/9789240083236
2. Masello, L., Castignani, G., Sheehan, B., Murphy, F., McDonnell, K.: On the road safety benefits of advanced driver assistance systems in different driving contexts. Transp. Res. Interdiscipl. Perspect. **15**, 100670 (2022)
3. Long, X., Nan, X.: Small and medium obstacle detection method in intelligent driving scenarios. Sci. Technol. Eng. **25**(09), 3778–3787 (2025)
4. Huang, Q., Hu, Z., Liu, H.: The influence of listening to signs on attention processing of visual warning information. Chin. J. Clin. Psychol. **30**(05), 1026–1031 (2022). https://doi.org/10.16128/j.cnki.1005-3611.2022.05.004
5. Zhao, H., Yang, W.: Vehicle audio recognition based on short-term energy and Mel cepstral coefficients. Sci. Technol. Eng. **18**(18), 197–201 (2018)
6. Parvin, H., Sarvi, M., Bagloee, S.A.: Detecting emergency vehicles with 1D-CNN using Fourier processed audio signals. Measurement **223**, 113784 (2023). https://doi.org/10.1016/j.measurement.2023.113784
7. Chen, H., Zhang, Z.: Hybrid neural network based on novel audio feature for vehicle type identification. Sci. Rep. **11**, 7648 (2021). https://doi.org/10.1038/s41598-021-87399-1
8. Ma, Q.: Noise suppression of sound signal transmission based on improved spectral subtraction. Electroacoust. Technol. **49**(03), 44–46 (2025). https://doi.org/10.16311/j.audioe.2025.03.015
9. He, Y., Bian, Q., Zhang, B.: Research on time domain feature analysis of vehicle audio signals. Electron. Prod. **7**, 61–63 (2021). https://doi.org/10.16589/j.cnki.cn11-3571/tn.2021.07.019
10. Mannem, K.R., Mengiste, E., Hasan, S., García de Soto, B., Sacks, R.: Smart audio signal classification for tracking of construction tasks. Autom. Constr. **165**, 105485 (2024). ISSN 0926-5805. https://doi.org/10.1016/j.autcon.2024.105485
11. Zhou, L., Li, Y., Yang, H., Bu, X.: Research on vehicle model recognition method based on Frost and MFCC features. Instrum. Technol. **6**, 58–63 (2023). https://doi.org/10.19432/j.cnki.issn1006-2394.2023.06.013
12. He, K., Zhang, X., Ren, S., Sun, J.: Deep residual learning for image recognition. In: Proceedings of IEEE Conference on Computer Vision and Pattern Recognition (CVPR), pp. 770–778. Las Vegas, NV, USA (2016)
13. Wang, J., Zhou, X.: Spectral classification and recognition method of asphalt mixture based on 1D-ResNet. Mod. Electron. Technol. **48**(08), 139–144 (2025). https://doi.org/10.16652/j.issn.1004-373x.2025.08.022
14. Golzari Oskouei, A., Balafar, M.A., Motamed, C.: RDEIC-LFW-DSS: ResNet-based deep embedded image clustering using local feature weighting and dynamic sample selection mechanism. Inf. Sci. **646** (2023). Art. No. 119374

MIX
Papier aus verantwortungsvollen Quellen
Paper from responsible sources
FSC® C105338

If you have any concerns about our products,
you can contact us on
ProductSafety@springernature.com

In case Publisher is established outside the EU,
the EU authorized representative is:
Springer Nature Customer Service Center GmbH
Europaplatz 3, 69115 Heidelberg, Germany

Printed by Libri Plureos GmbH
in Hamburg, Germany